# PRACTICAL FIELD ROBOTICS

# PRACTICAL FIELD ROBOTICS

## A SYSTEMS APPROACH

**Robert H. Sturges, Jr**
*Virginia Tech, USA*

WILEY

This edition first published 2015


*Registered Office*
John Wiley & Sons, Ltd, The Atrium, Southern Gate, Chichester, West Sussex, PO19 8SQ, United Kingdom

For details of our global editorial offices, for customer services and for information about how to apply for permission to reuse the copyright material in this book please see our website at www.wiley.com.



*Library of Congress Cataloging-in-Publication Data*

Sturges, Robert H., Jr.
Practical field robotics : a systems approach / Robert H Sturges, Jr.
pages cm
Includes bibliographical references and index.
ISBN 978-1-118-94114-0 (cloth)
1. Mobile robots. I. Title.
TJ211.415.S78 2015
629.8′932–dc23

2014030550

A catalogue record for this book is available from the British Library.

Set in 10/13pt Times by SPi Publisher Services, Pondicherry, India

1 2015

# Contents

**Preface** ix

**1 Overview of Field Robotics** 1
1.1 Introduction 1
1.2 Methodology 3
1.3 High-Level Decisions 3
Problems 4
Notes 4

**2 A Mobile Robot System for Nuclear Service** 7
2.1 Field Environment: Commercial Nuclear Plants 7
2.2 Field Work: Component Maintenance 8
2.3 Equipment Requirements 9
2.4 Conceptual and Operational Designs 12
2.5 Safety and Reliability 19
2.6 Detail Designs of the Service Arm 20
2.7 Detail Designs of the Walker 20
2.8 Conclusion 21
Problems 21
Notes 22

**3 The Largest Mobile Robot in the World** 23
3.1 Field Environment: Underground Mining 23
3.2 Field Work: Continuous Coal Haulage 25
3.3 Equipment Requirements 26
3.4 Conceptual and Operational Designs 29
3.5 Safety and Reliability 30
3.6 Detail Conceptual Designs 30
3.7 Conclusion 31
Problems 31
Note 31

4 **A Mobile Robot for Mowing a Lawn** 33
4.1 Field Environment: Suburban Lawns 33
4.2 Field Work: Navigation and Mowing 34
4.3 Equipment Requirements 34
4.4 Conceptual and Operational Designs 35
4.5 Safety and Reliability 37
4.6 Detail Conceptual Designs 37
4.7 High-Level Decisions 37
4.8 Conceptual Design—Technologies 38
4.9 Conceptual Design—Set Parameters 40
4.10 Conceptual Design—Operate Robot 42
Problems 42
Notes 43

5 **The Next Levels of Functional Detail** 45
5.1 Quantifying Conceptual Design 45
5.2 Quantifying *Send Sound* 46
5.3 Quantifying *Receive Sound* 53
5.4 Quantifying *Interpret Sound* 56
5.5 Design Choices—Setting Parameters 65
5.6 *Select* a *Platform* 66
5.7 *Select Frequencies* 68
5.8 *Select Motions* 70
Problems 72
Notes 72

6 **Operate Robot** 73
6.1 *Control System* 75
6.2 *Control System Select Operation* 76
6.3 All About `main()` 78
6.4 *Control System—Control Motions* 79
6.5 *Control Motions—Rotate Motors* 81
6.6 *Control Motions—Design Infrastructure* 83
6.7 *Control Motions—Program Speeds* 88
6.8 *Control Motions—Move Robot* 89
6.9 *Control Motions—Sequence Motions* 92
6.10 *Control Information* 92
Problems 102
Notes 103

7 **Software Functions** 105
7.1 Displays: To Place Needed Information to the User Screen 107
7.2 Field Data and Triangulation: Geometric Locating Functions 109
7.3 Operation: The Calls that Make the Robot Move and Stop 121
7.4 History and Diagnostics: The Immediate Past Used for Analysis 130
Problems 136
Note 137

**Appendix A: Myth and Creativity in Conceptual Design** 139

**Appendix B: Real-World Automation Control through the USB Interface** 159

**Appendix C: Microchip Code for USB Board to PPM Translation** 173

**Appendix D: Selected Electronic Parts for Mowing Robot** 179

**Appendix E: Software Concordance** 181

**Appendix F: Solutions** 187

**Index** 197

# Preface

This is an introductory book to the area of Field Robotics. Since the number of practical examples is very large, we can only hope to introduce a few examples in detail. We will approach the subject with a systems design methodology, showing the reader every important decision made in the process of planning, designing, making, and testing a field robot. The book will cover electronic, electrical, mechanical, control, and software disciplines as needed. We begin with discussions of industrial field robots and then delve into the details of a practical machine that can mow your lawn. Unlike commercial devices that mow random patterns, this one "knows where it is." The robot is also cost effective so that the serious reader may follow the logic, understand the steps, and carry out its construction.

Our intention is to instruct the reader on "how" and "why" things get done the way they do, and we aim the book at advanced high school students, college students, serious hobbyists, and engineers in practice. We provide the planning tools and present either summaries or detailed descriptions of the working parts of a successful machine. For the lawn mower, we ask you to keep in mind that such a machine is not a toy and can be potentially dangerous to construct and use. Caution and safety glasses are a must. The software is listed in full, and we expect that it will be modified to suit as design changes are considered and implemented. Our engineering decisions may not be yours and we encourage variation, experimentation, and extensions, especially as technology advances. The systematic design approach we use will make this effort highly visible and useful for others as well.

On the personal side, my interest in field robotics goes back to the 1960s when there were very few examples. Teleoperators (man-in-the-loop systems) had just become servo-operated and the door was opened for computer-assisted manipulation. My colleagues and I at the Draper Labs at M.I.T. began studying the earliest assembly systems, while the industrial world saw one of the first "made for work" programmable manipulators, the hydraulic Unimate™. From these pick-and-place systems, grew the merging of sensors with actuators at M.I.T.'s Artificial Intelligence Lab. At about that time also, university labs began experimenting with "less than real-time" mobile robots since computers were still very limited in memory and speed.

Decades later came the appearance of more sophisticated systems, and I have selected several of these "pet projects" from among the many I have had the pleasure to have developed along with stellar groups of colleagues. The lawn-mowing robot, to which I devote the most detail here, was inspired by similar discussions and, for good or ill, is my own. As many of these systems exist for their usefulness as well as their potential, the advent of ever-accelerating technology places more and more ideas into the realm of practicality. As we go to press there are hundreds of such projects in the works and in the field, airborne, and underwater systems becoming prominent. It is my hope that this volume will inspire you to go beyond discussion and into the field with me.

# 1

# Overview of Field Robotics

## 1.1 Introduction

Practical Field Robotics comprises the design and fabrication of machines that do useful work on their own, for the most part. Field Robotics separates us from robotics done in a protected laboratory environment. It is also somewhat removed from theoretical robotics that underpins much of what we do, but may not employ in the realization of machines. Laboratory machines may walk like a human[1], or simply perform useful tasks that humans do in other ways.

When asked to describe a "robotic dishwasher," new students often elaborate on a machine that picks up one dish at a time and does the washing the same way a human would. This approach ignores the success of pumping scalding water around in a sealed box. The former approach expresses limitations that were never imposed by the question.

The robotics literature abounds with examples that expressly imitate how humans are built, rather than what they aim to do. Since our hands and eyes evolved to allow us to swing from trees and pick fruit, they may not afford the best characteristics needed for a more modern task[2]. We should not be tempted to imitate and then automate things we ought not do in the first place. For example, Elias Howe has been quoted as saying that he was inspired by watching his wife sewing in his work to invent the sewing machine[3]. Analysis of how people sew and how his machine sews shows that this could not be the case. Our methodology will steer us away from these pitfalls.

A successful design should always consider such constraints as artificial and aim for the *function* to be performed[4]. We will see (Figure 1.1) an expression of high-level functions that not only informed the development of our case studies, but also suggested the layout of the chapters in this book.

We will explore three examples of systematically designed field robotic systems, each illustrating key points of the design procedure and important lessons learned in the field. The first

*Practical Field Robotics: A Systems Approach*, First Edition. Robert H. Sturges, Jr.

Companion Website: www.wiley.com/go/sturges

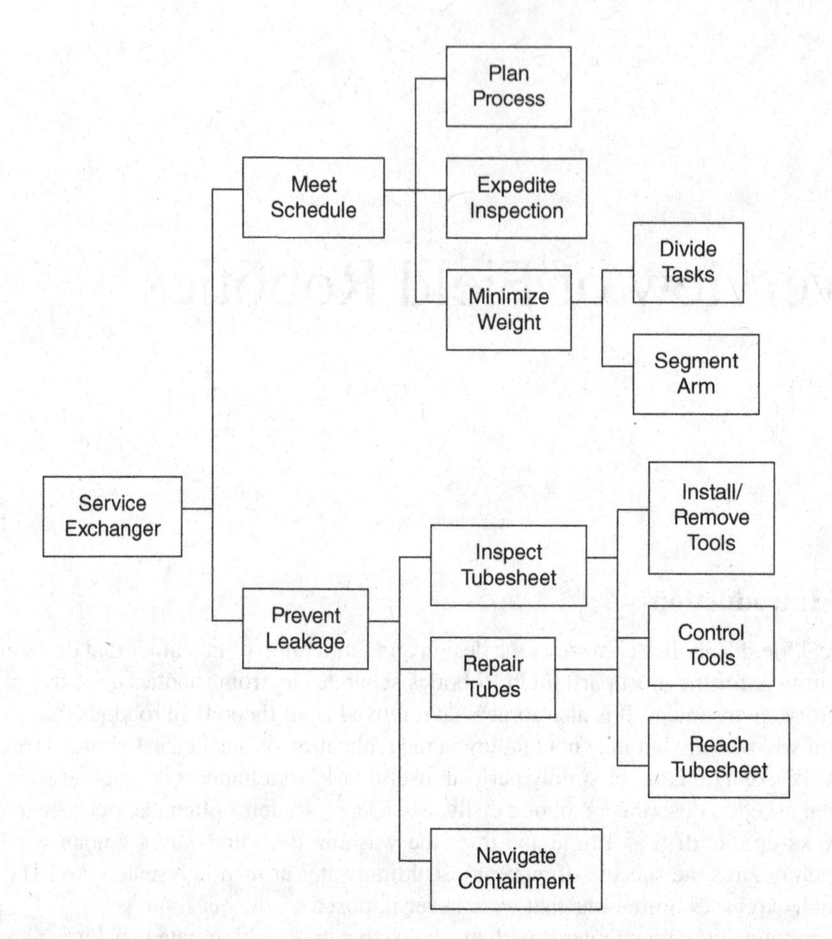


**Figure 1.1** The beginning of the FBD for a practical field robot for nuclear service

example is a mobile robot system, actually a pair of cooperating robots, used for field repair work in the commercial nuclear power plant area. This system was developed over time by several design teams considering the functions to be performed, and we will illustrate them here. Details of the system design remain proprietary but we can address the higher-level decision-making processes by reference to a patent issued to the author[5]. We will see that fully automatic control was not employed, but rather the concept of *teleoperation*, in which there are personnel in-the-loop at all times[6,7]. Special-purpose equipment and tooling were employed to create a level of autonomy for some subsystem tasks.

The second example involves the design and operation of the largest autonomous mobile robot ever built, to our knowledge. Its mission is to haul coal from an underground mine. Its field versions weigh hundreds of tons and span lengths of 160 m from end to end. Briefly, it comprises a continuous miner at the head end, a number of linked conveyors, and telescoping tail piece. While one terminal end does need a human driver, the balance of the system is autonomously driven through the mine, needing only periodic observation of its segments to ensure that *no personnel* come within its vicinity.

Finally, we present a detailed account of the design process and the operation of a low-budget mobile robot for automatically mowing a lawn at an affordable cost. To our knowledge this has not successfully been done until now. We initially examined six approaches to this problem, and each will be discussed before delving into the details of the selected concept.

## 1.2 Methodology

For all of the design examples, a common tool will be employed: the process of Value Engineering. The literature is replete with many examples of its use in both military and commercial designs[4]. We have included Appendix A to explain the process in detail, but will touch upon its major themes in this introduction. In summary, it expresses functions in a structured way such that the *intent* of the designer is made clear to all parties to the exercise. We do this by expressing each function of our design at successively higher levels of detail as we scan a page from left to right. The graph created by our consideration of intent and challenges is called a function block diagram (FBD). Also as we scan the page, the rightmost functions answer the question "how," and the leftmost answer the question "why."

Each function, in turn, is expressed with a single active verb and a single measurable noun, placed in a little box. A decision or artifact is shown in a rounded-corner box. Figure 1.1 shows a part of such an expression of the high-level function *Service Exchanger,* along with successively higher levels of detail in a tree-like hierarchy. The arcs connecting each function block can be read as "and." To give more meaning to the terms in use, the system to be discussed in the next chapter involves the servicing of the primary heat exchangers in a power-generating installation, but not the reactor itself. As mentioned, the rubrics of Value Engineering represent functions with verb–noun combinations. Attached to these may be *allocations* that specify quantitative constraints, which are not functions. Thus, in order to *Service Exchanger* we need to do two things: *Meet Schedule* and *Prevent Leakage.* These may seem obvious in hindsight, but functions rarely are. These early decisions have the most profound effects on the design details and express our unique choices. We emphasize that there is no single "correct" function diagram, but that the design team agrees that it is valid.

As we will discover, from the more detailed functions at the extreme right-hand end of the abbreviated FBD, we could have chosen to avoid any teleoperation and decide on the aggressive development of more "robotic" approaches. We put "robotic" in quotes to distinguish this enterprise of Field Robotics as the realization of a more intelligent automated process. We employ the term "practical" to mean that what we design and build must work and do so according to constraints. While we may slip into some philosophical discussions along the way, the leading emphasis is on practicality in terms of functionality and limited cost. In fact, Value Analysis was originally conceived as a cost-containing measure, but its adoption easily leads to innovation as well.

## 1.3 High-Level Decisions

We should mention here that function logic, the means by which we reason our way from general to specific, is not normally encountered in the iterative cycle of "analysis, synthesis and evaluation"[8]. It coexists with our "normal" way of thinking, and offers a great deal in terms of organization and expression of our designs to others. A functional map of a product or process is far easier to understand by those on the "design team" with separate interests and other skill-sets, than would be a stack of engineering drawings. Moreover, we can use it to innovate by simply supposing that a certain function is restricted, prescribed, or dependent on a specific technology that should be avoided.

The adoption of hierarchical functions to describe the intent of the design also serves to guide the discussion of the design and implementation. Rather than present a detailed description of what each portion of the robot *is,* we will explain what each of its functions *do.* The order in which this will occur follows the function diagram itself, beginning at a high level of abstraction, and gradually exposing more and more detail. In this way, the reader is not faced with synthesizing an entire system in his/her mind, but can "drop in" at any desired level of detail to learn *why* and *how* the robot was conceptualized and realized.

The practical decisions expressed in Figure 1.1 will guide us through the ideation and realization of practical examples of field robotics. Unlike Elton John, who sang[9],

> And each day I learn just a little bit more.
> I don't know why, but I do know what for.

we *will* know precisely "why" a choice was made by simply looking leftward on the page. We will also know "how" by looking to the right. Unlike a flow chart, common to business and software planning, we will have an expression of the *raison d'être* rather than just a prescribed sequence. Incidentally, a vertical reading of any column of functions will also express, roughly, an operational sequence too[10]. In addition, the functional method also keeps us from falling into familiar patterns of thought. For example, the phenomenon of *Einstellung*[11], choosing a familiar good idea and neglecting unfamiliar better ideas, is reduced since the entire design team is empowered to contribute to the high-level functions, at least.

## Problems

1.1 Consider the manual task of sorting mail into letters, flats, and packages. How might this be done robotically?

1.2 Describe at least one teleoperator now popular with consumers.

1.3 Please describe the systems used in "self-driving" automobiles.

1.4 Reviewing Figure 1.1 and Appendix A, please create an FBD for the functions we perform to go from sleeping to arriving at work/school.

## Notes

1. Kanabe, C., Hopkins, M. and Hong, D. (2012) Team CHARLI: RoboCup 2012 Humanoid AdultSize League Winner, in *RoboCup 2012, Lecture Notes in Computer Science* (eds X. Chen, P. Stone, L.E. Sucar and T. Van der Zant), Springer, pp. 59–64.
2. Jacobsen, S.C. *et al.* (1986) Design of the Utah/M.I.T. Dexterous Hand. Proceedings of the IEEE Conference on Robotics and Automation, April 1986, Vol. 3, pp. 1520–1532.
3. http://en.wikipedia.org/wiki/Elias_Howe (accessed June 12, 2014).
4. Fowler, T.C. (1990) *Value Analysis in Design*, Van Nostrand Reinhold.
5. Westinghouse Electric (1979) US Patent 4,168,782, September 25, 1979.
6. Thring, M.W. (1983) *Robots and Telechirs*, Halsted Press.

7. Corliss, W. and Johnson, E. (1968) Teleoperator Controls, NASA SP-5070, Lib Cong Cat #73-600664.
8. http://en.wikipedia.org/wiki/Bloom's_Taxonomy (accessed June 20, 2014), especially Cognitive Stages 1.4, 1.5 and 1.6.
9. http://www.eltonography.com/songs/this_song_has_no_title.html (accessed June 24, 2014).
10. Sturges, R.H., O'Shaughnessy, K. and Reed, R.G. (1993) A systematic approach to conceptual design based on function logic. *International Journal of Concurrent Engineering: Research & Applications (CERA),* **1**(2), 93–106.
11. Bilalic, M., McLeod, P. and Gobet, F. (2008) Why good thoughts block better ones: the mechanism of the pernicious Einstellung (set) effect. *Cognition*, **108**(3), 652–661.

# 2

# A Mobile Robot System for Nuclear Service

## 2.1 Field Environment: Commercial Nuclear Plants

The field of nuclear service presents a wholly distinct set of constraints from ordinary laboratory robot development. The interior of a commercial nuclear plant spans the gamut from "shirt sleeve" operations to highly restricted areas with 100% humidity, temperatures over 100 °F, and radiation levels that can cause harm to humans at exposures of only 15 min. (Semi-skilled maintenance personnel wear dosimeter badges that monitor the monthly "dose" of ionizing radiation.) Further, plant workers generally need to carry equipment through a maze of passages and to negotiate narrow steps and ladders. This plant layout and vertical barriers restricts the worker to carrying no more than 16 kg at a time.

Generally, no guide-posts exist to assist in navigation from the "safe" region of the containment building to the steam generator. (At least one containment facility was equipped with overhead rails for ferrying equipment, but this was an exception.) A mobile robot would need very adaptable code to recognize and navigate its location. As mentioned, the pathway to the steam generator may include vertical segments, making mobile robotics even more unlikely.

The location of the entrance to the steam generator is sealed with a 10-cm-thick hatch that is bolted into place. The location of the bolts is not precisely known with respect to any common reference, making automated removal by fixed programming an unlikely choice. This hatch, when removed, uncovers the lower part of the steam generator that ducts the high pressure, high temperature water into the heat exchanger tubes. Thus, one section of this workspace, referred to as the "channel head," is devoted to a large descending water pipe of about 1 m in diameter. The ceiling of the channel head (the tubesheet) is studded with several thousand tube ends, each approximately 20 mm in diameter, as shown in Figure 2.1. The locations of these tube ends are precise relative to each other (about 0.1 mm) but very imprecise with respect to the opening to the channel head uncovered by the hatch.

*Practical Field Robotics: A Systems Approach*, First Edition. Robert H. Sturges, Jr.

Companion Website: www.wiley.com/go/sturges

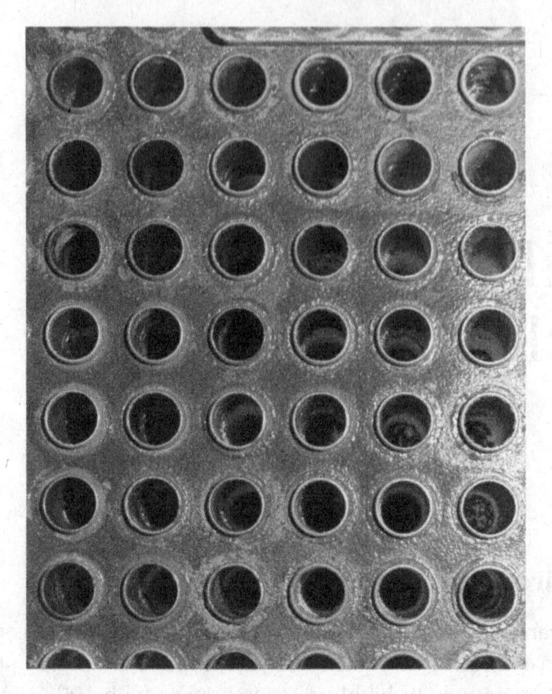

**Figure 2.1** A portion of a tubesheet model with 20 mm diameter tube ends (Source: Reproduced with permission. © Westinghouse Electric Company LLC)

Again a mobile robot would need very adaptable code with many degrees of freedom to recognize and navigate its way to the tubesheet.

At the time of this system's design and introduction, algorithms for general-purpose localization in six degrees of freedom were not yet available, even in a laboratory environment. The choice for moving the service equipment to the point of use was imposed by the field environment and the work itself: people were needed to carry the equipment, and people were needed to (at least) direct the work from a remote location.

## 2.2 Field Work: Component Maintenance

Periodic re-fueling of a commercial nuclear reactor is beyond the scope of this work, but it is carried out with assistance from programmable machines that must be installed onto the reactor vessel and work with built-in plant equipment for handling reactor components. The time needed to perform a re-fueling must be minimized, since the cost of not producing power may exceed US$1 million per day. For this reason periodic maintenance of other reactor components, for example the steam generators in a pressurized water reactor (PWR), may be restricted to this time interval and calendar constraints.

As an example, the heat exchanger tubing must be inspected for leakage, however slight, and steps must be taken to re-seal or plug the suspected tube under the field conditions mentioned above. Since a steam generator of the type considered here uses thousands of such tubes, the net effects on facility performance are very small. In any event the inspections must be carried out with high reliability in the prescribed window of time with equipment that must withstand the rigors of the field environment and feature such portability that semi-skilled personnel can carry, locate, and assemble the equipment in 16-kg units.

## 2.3 Equipment Requirements

There are many constraints on materials used in a reactor building. The use of halogens or materials containing halogens is restricted. This means that ordinary hydraulic oils are forbidden. Polyalkyene glycol is acceptable but requires special elastomeric hoses to carry it without causing rapid degradation. Any device deployed in a steam generator channel head must also use a small range of approved materials. For this brief discussion, alloys of aluminum and stainless steel are acceptable, but lead (Pb) is not. The chemistry of the water is constantly monitored, so that accidentally leaving any tools or equipment behind would quickly be found and give cause for a possible unplanned shut-down of the facility.

There have been many distinct types of equipment designed and utilized in Nuclear Service, and we will describe the design and construction of one type, namely a pair of teleoperated robots. A pair had been the choice of designers since the practical application of field work could be divided into two stages[1]. A service "arm" could be used to maneuver a "tubesheet walker" into position for its field work. The deployed arm is shown in Figure 2.2. It could also maneuver specially constructed tools to service the walker, in turn. The walker would suspend itself from the ceiling (or tubesheet) of the vertically mounted heat exchanger and navigate across its tube-studded surface, as shown in Figure 2.3. Ancillary equipment needed in the chamber surrounding both robots comprised lights and video cameras to obtain a coarse but reliable view of the operations.

**Figure 2.2** Arm deployed in channel head (Source: Reproduced with permission. © Westinghouse Electric Company LLC)

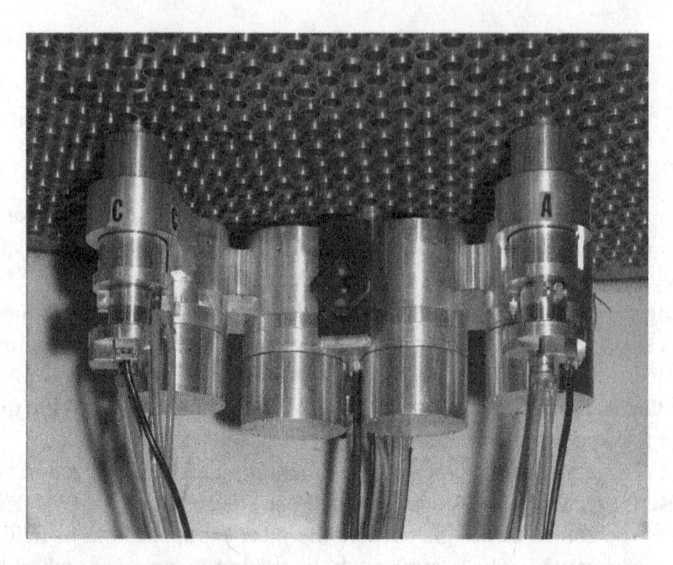

**Figure 2.3** Walker deployed on tubesheet of steam generator (Source: Reproduced with permission. © Westinghouse Electric Company LLC)

**Figure 2.4** Arm installation track in a model channel head (Source: Reproduced with permission. © Westinghouse Electric Company LLC)

In addition to the 16 kg limitation, the entry passage into the chamber where the inspection and possible maintenance would occur consisted of a single 406 mm "manway" through a wall of about 30 cm of stainless steel-cladded carbon steel. (The breadth of one's shoulders certainly exceeds 406 mm, so the procedure for entering and leaving a channel head comprised adopting a "diving" posture, with the arms tightly pointed with the long-axis of the body.) This geometric constraint led to the addition of a guide track for installing the arm, as shown in Figure 2.4.

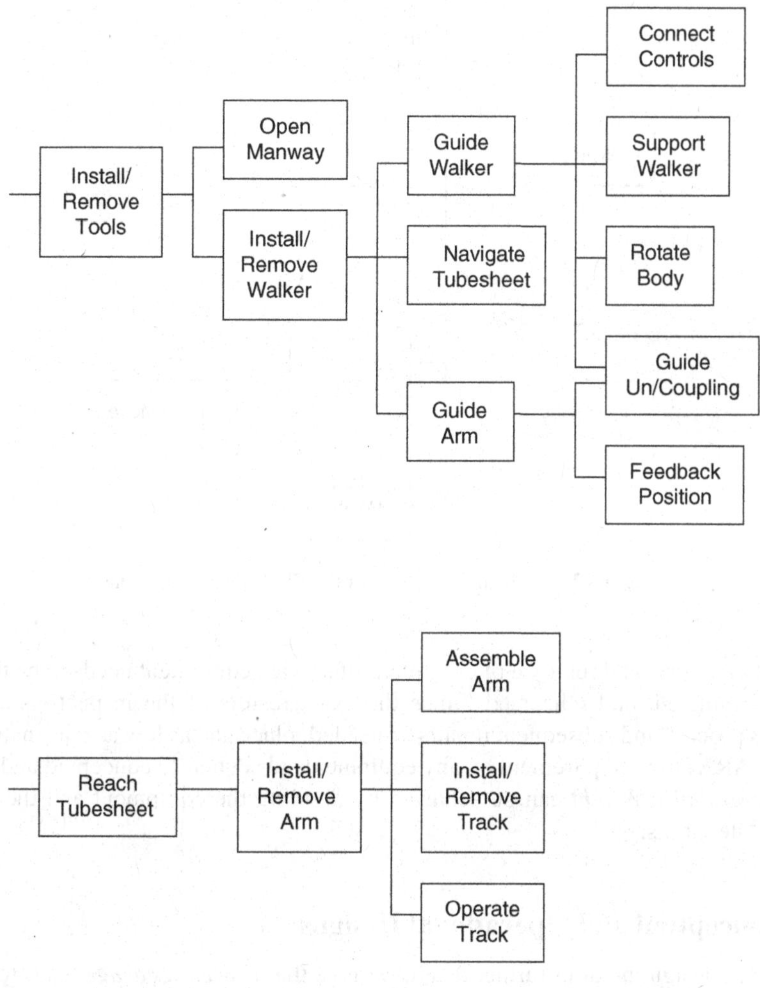


**Figure 2.5** The second portion of the FBD for the robot pair

The remaining access profile would need to accommodate the walker during its transition into position on the tubesheet. These numerical details are beyond the scope of our functional discussion, but we will describe some of them as needed to explain some of the design choices made.

A function block diagram (FBD) of the first stages in the engineering of these robots is given in Figure 1.1, Figure 2.5, and Figure 2.6. At each branching of the diagram design decisions were made based on the known constraints and the "how–why" strategy of the FBD being built. For example, a functional starting point was selected as *Service Exchanger*. The adjective "heat" would be attached to the function block for a detailed identification if needed. In order to *Service Exchanger*, at least two functions are needed: *Meet Schedule* and *Prevent Leakage*. How these are expressed functionally follows. To satisfy the small window of time in which to *Prevent Leakage, Meet Schedule* needs at least three subfunctions: *Plan Process, Expedite Inspection,* and *Minimize Weight.* In practice, every step of

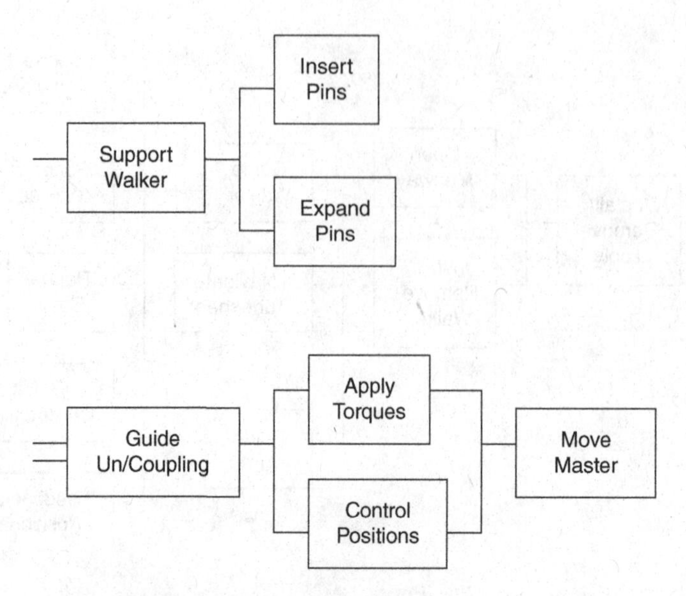


**Figure 2.6** The third portion of the FBD for the robot pair

the installation, use, and removal of each piece of service equipment needs to be thoroughly planned, simulated, and rehearsed. Since the exact results of the inspection will not be known, its process and subsequent repairs, if needed, place an "as low as reasonably achievable" (ALARA) time requirement on any equipment subsequently concepted and designed. Similarly *Minimize Weight* can be achieved by dividing the equipment and the tasks into easily handled units.

## 2.4 Conceptual and Operational Designs

Building our design intent in further detail, we note that *Prevent Leakage* leads to functions that have already been hinted at, namely *Navigate Containment, Inspect Tubesheet,* and *Repair Tubes.* A decision to not automate the first function leads to manual haulage of all equipment from the "safe" part of the containment building to near the point of use at the channel head. Thus, one can see that, even at this high level, key decisions are explicitly shown in the FBD, and are therefore subject to the scrutiny of many engineers, managers, and operators. However, we do not begin conceptual design (beyond the existence of some "arm") until we have gone deeper in terms of functionality and obtain agreement among all concerned parties.

Several necessary functions appear connected to *Inspect Tubesheet,* and *Repair Tubes.* In order to *Inspect Tubesheet,* and *Repair Tubes,* the equipment must at least *Install* and *Remove* its *Tools, Control* these *Tools,* and explicitly *Reach* the *Tubesheet.* This latter function is not obvious because the shape of the channel head is a quarter sphere, and reaching the most peripheral tubes in the tubesheet presents challenging design issues. Figure 2.7 shows this requirement schematically. The interconnected "and" arcs in the FBD indicate that to either *Inspect Tubesheet* or *Repair Tubes,* the three dependent functions must be carried out.

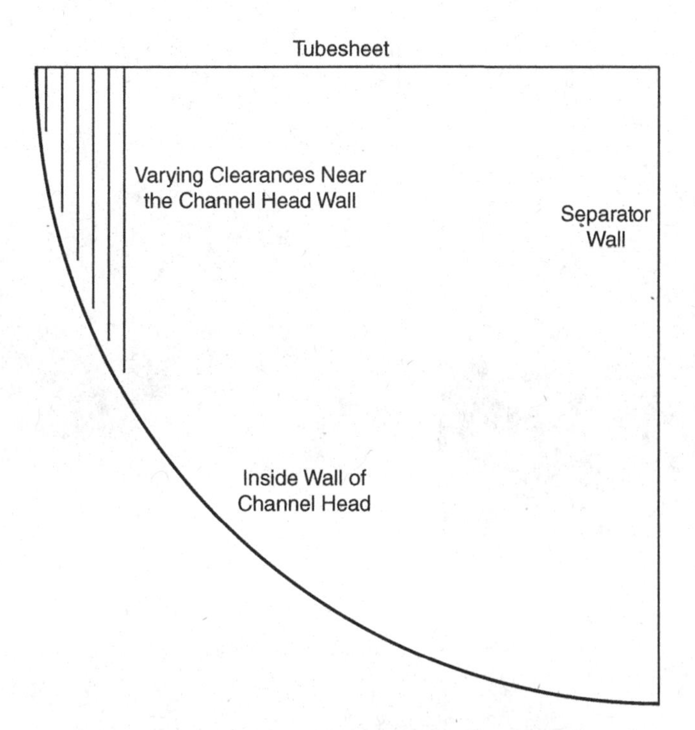


**Figure 2.7** Geometric constraints at the edges of the tubesheet

A strictly hierarchical arrangement of functions could, of course, be constructed, but several identical functions would be repeated.

Referring to Figure 2.5, two distinct portions are shown that further expand the functions from the left-most panel of the FBD in Figure 1.1. The *Reach Tubesheet* function is subdivided into geometrical challenges not explicitly called out functionally, plus the function *Install/Remove Arm.* This function will be further decomposed in light of the equipment requirements noted above. The intent of the design will be to carry out three more subfunctions, namely *Assemble Arm, Install/Remove Track,* and *Operate Track.* Figure 2.8a–c shows three parts of the arm unassembled, while Figure 2.9 shows the assembled arm on its installation track. Recall that Figure 2.4 showed the track alone fitted into the restricted space of a manway.

In order to *Install/Remove Tools,* a major conceptual decision is depicted functionally. The arm itself will not perform the tight-tolerance work on the tubesheet, but it will be assisted by a mobile robot that "walks" on the tubesheet of the channel head. The reasons for this decision recall the equipment requirements of the inaccuracy of the manway location and the high accuracy of the tube end locations themselves. Thus, a walker is envisioned in the subfunction *Install/Remove Walker.* The new function *Navigate Tubesheet* is now part of the work that the walker must do: handling the tools. The function of *Open Manway* is not shown, but it is done manually. A number of supporting tools, also not shown, accomplish this with semi-skilled workers. One may certainly consider a large number of alternatives at this juncture. The FBD shown is the result of much discussion and conceptual brainstorming. It may have instead transpired very differently than it did. Recall that an FBD is not unique and needs only to be valid in the minds of the many team members creating it.

**Figure 2.8** Separated portions of the arm: (a) part 1; (b) part 2; and (c) part 4 (Source: Reproduced with permission. © Westinghouse Electric Company LLC)

As we ask "how" to *Install/Remove Walker,* we determine that at least two functions must be performed: *Guide Walker* (to the tubesheet from the manway) and *Guide Arm,* the agent that would perform both that function as well as bringing and removing tools from the walker. The action of delivering/removing a tool between the arm and the walker is shown in Figure 2.10. This photograph certainly puts the design cart before the functional horse, but may aid in understanding the intent of this process at this point.

How can we conceptually *Guide* the *Walker* and the *Arm*? Fully autonomous control is ruled out by the Equipment Requirements, and the lack of technology (at that time) to accomplish it. The answer is not yet explicit since many other functions intervene. To *Guide Arm* at least two functions are needed: *Guide Un/Coupling* and *Feedback Position.* In this latter

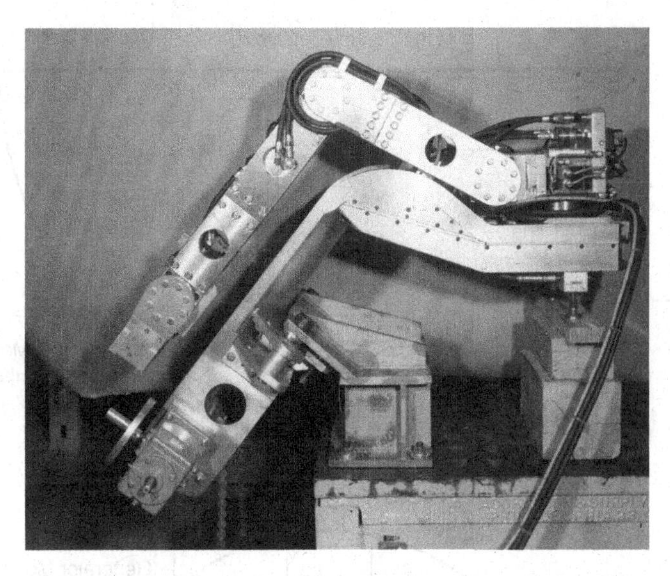

**Figure 2.9** The fully assembled arm (Source: Reproduced with permission. © Westinghouse Electric Company LLC)

**Figure 2.10** The arm handing-off an inspection tool to the walker (Source: Reproduced with permission. © Westinghouse Electric Company LLC)

function, the word "feedback" is used as a verb, since some loop closure is needed to accommodate for the inaccuracies of positioning in the manway. For the walker, at least four functions are required: *Connect Controls* (actually non-obvious), *Support Walker, Rotate Body* (of the walker), and *Guide Un/Coupling*.

The *Connect Controls* may seem to be an obvious thing to do at this point, but there is an anecdote from field experience. Both the walker and the arm are connected to the "safe" area where the Service Exchanger system controls are located by "umbilical" cables. Figure 2.11

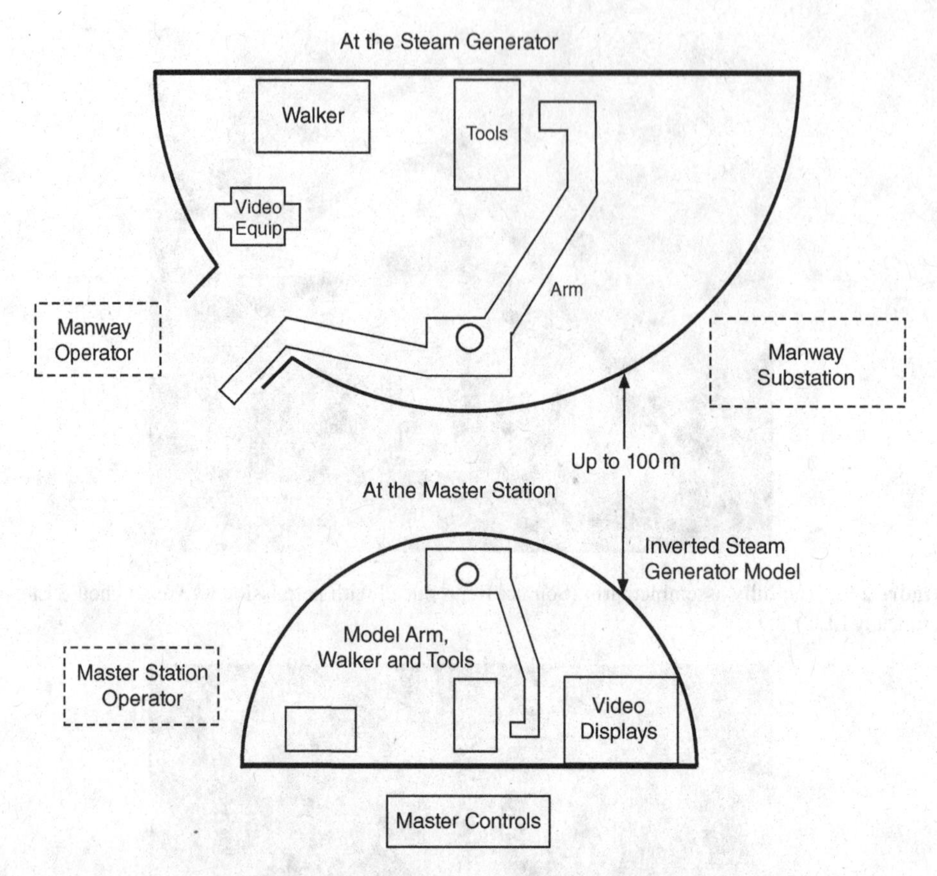


**Figure 2.11** Schematic of the service system

shows this arrangement as about 100 m apart. In performing the assembly of these cables, one of the set was accidentally exchanged for another, and could not be mated to its counterpart. The workman was undaunted: he grabbed his diagonal cutters and removed the connector pins that prevented the linkage and "completed" his task. It was many hours later that the cause of the control system power-on test failure was traced to what appeared to be an intact joining of essential cables.

Figure 2.6 depicts the right-hand side of this conceptual FBD, wherein two of the higher-level functions are decomposed into lower-level ones that lead more directly to design parameters. The *Support Walker* function is needed to indicate that there must be devices or structures that will retain the walker against gravity while navigating the tubesheet. Design choices may be numerous, but the principle exploited was *Insert Pins* and *Expand Pins.* The open ends of the tubes themselves are used to support the walker through the use of expandable mandrels. This choice is supported by the knowledge that the locations of the tube ends are within close tolerances. The number of pins/mandrels remains to be determined, as well as their method of positioning beneath the tube ends. Our abbreviated FBD in Figure 2.6 leaves off at this point, but it will be understood that many more detailed conceptual decisions had been made leading to the realization of the walker itself. We show a "top side" view of

**Figure 2.12** The walker as seen from the top (Source: Reproduced with permission. © Westinghouse Electric Company LLC)

the walker in Figure 2.12. It is supported in the tubesheet and in the process of receiving a tool from the arm in Figure 2.10.

A principal task in order to *Install/Remove* Tools is to perform a hand-off by coupling a tool grasped by the end of the arm to a mating face of the walker. Before detailing the functions of the coupler itself, we need to first indicate how to *Guide Un/Coupling* between the arm and the walker. Our FBD clearly shows a common "and" connection between the arm and the walker (as well as the tools used). This guidance requires high-level design choices, which were *Apply Torques* and *Control Positions* of both the walker and the arm. Torques are needed to locate the end of the arm against gravity in the channel head. A different set of torques is needed to rotate the walker into position so that the coupler parts can engage. Figure 2.13 shows a feature built into the walker controls for performing such alignment[2,3]. It consists of a three-axis servo (x, y, and iris) driven by a focused light beam produced by the walker and received by the tool or arm. It serves to precisely locate both mating parts of the coupler while under manual control of both the arm and the walker.

The question of how to accomplish this "last" function appears at the right-hand end of the FBD in Figure 2.6. A key design choice was made to operate this pair of robots in a teleoperator mode using scaled versions of both. In this case the operating volume of the channel head needed to be reproduced in scale wherein both the walker and arm were manually manipulated to *Apply Torques* and *Control Positions.* Figure 2.14 shows an engineer operating a scaled model of the "master" walker while obtaining direct visual feedback from a video system located inside the channel head. Notice that all of the tube ends are represented, but that the model channel head is inverted with respect to gravity. This choice was made to reduce fatigue on the operator at the expense of not realistically modeling umbilical cables feeding the walker. A similar "master" for the arm is mounted in the scaled channel head to provide the operator with a relatively good representation of the arm location at all times. This "model mode" of control drives each joint on the arm and walker in a one-to-one servo mode, avoiding the use of digital feedback and the relatively primitive personal computers available at that time.

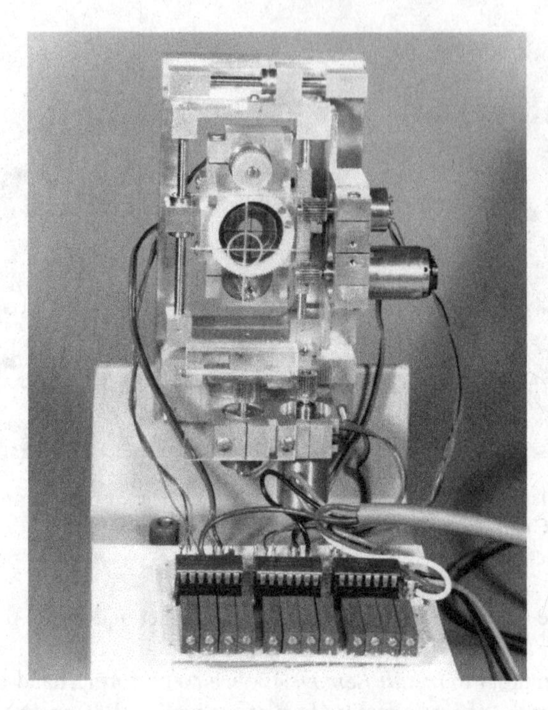

Figure 2.13 Docking display device (Source: Reproduced with permission. © Westinghouse Electric Company LLC)

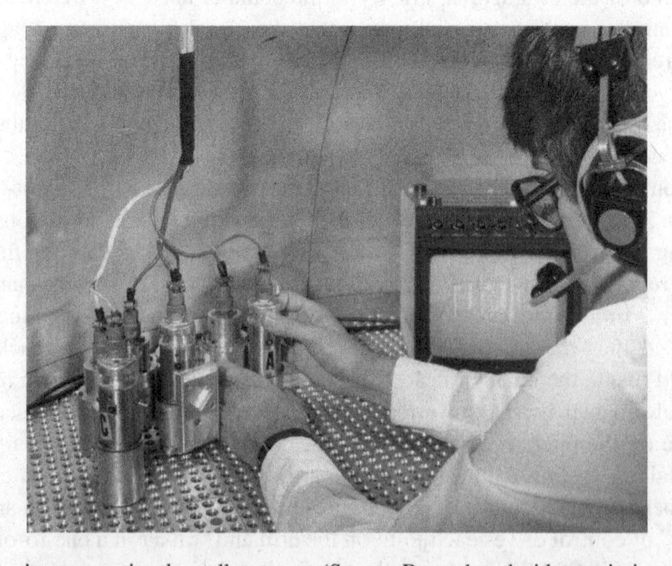

Figure 2.14 Engineer operating the walker master (Source: Reproduced with permission. © Westinghouse Electric Company LLC)

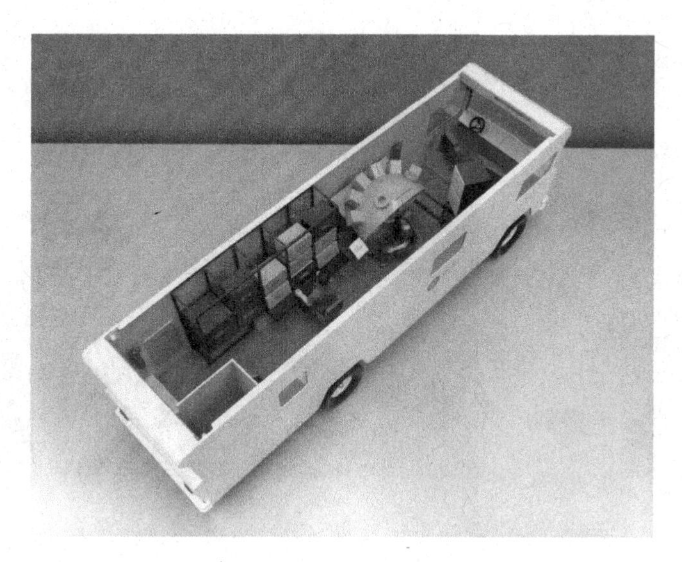

**Figure 2.15** Scale model of mobile service system (Source: Reproduced with permission. © Westinghouse Electric Company LLC)

## 2.5 Safety and Reliability

Conditions for Safety and Reliability were guided at a high level with the ALARA requirement. Three areas were subjected to Failure Modes and Effects Analysis (FMEA)[4]: the plant components; the repair team; and the repair equipment. Each of the tasks to be performed on plant components was subjected to approval by the plant operators and the Nuclear Regulatory Commission. These processes are beyond our scope. The repair team similarly was constrained by well-established safety and radiation dosage requirements, as mentioned above. In addition, the control module location (the robot and tooling controls) were designed for set-up in a "shirt-sleeve" environment in the containment building, as shown in Figure 2.11 schematically. The control units were designed and built with self-checks and spare modules. Operations were carried out remotely at half-scale with video windows and docking instrumentation. Safety to personnel was enhanced by installing a cover over the open feedwater pipe in the channel head as soon as the hatch was removed and stored. After a set of cameras and lights were attached to this cover, the arm mounting track was installed in the manway, preventing access to anyone while operations were carried out. Using the remote controls in "model mode" reduced the need for personnel to enter the relatively hot, humid, and radiating environment of the channel head.

When the system proved to be effective, greater umbilical lengths were designed. Plans were made for housing all of the control gear and tooling in a dedicated trailer so that the reliability of the two-robot system would be enhanced, and that the readiness of the entire operation could be practically made "24/7" (Figure 2.15). A more general discussion of safety in teleoperation is found in Sturges (1992)[5]. For removal of any residual irradiated material adhering to the arm or the walker, both were designed with water-tight seals so that they could be chemically washed down after each use. This operation occurred before the equipment was removed from containment after each use.

**Figure 2.16** The arm coupler, a coupler block, and the coupler control (Source: Reproduced with permission. © Westinghouse Electric Company LLC)

## 2.6 Detail Designs of the Service Arm

Referring to Figure 2.8, we see the arm in its several parts in order to satisfy the weight-per-unit constraint. When complete, the arm comprised four degrees of freedom: an azimuth rotation, followed by three rotational joints. Fully assembled, we see the arm comprising a number of hydraulic hoses and couplings (Figure 2.9). The hydraulic power unit is external to the channel head and is not shown, but an electrical umbilical leads from the base of the arm to the remote control area. The end of the arm carries the movable part of a coupler for engaging both the walker for its installation and the tooling used for inspection and service (Figure 2.16). While the mounting of the arm loading track could not be perfectly aligned with the plane of the tubesheet, the coupler components assured a positive lock on all tooling.

The master control model for the arm comprises a series of resistive servo potentiometers connected to the hydraulic actuators of the arm. These provide feedback to the master arm control of the four rotational axes. The controller also self-checks each connection from the arm master to the operating arm to ensure reliable control.

## 2.7 Detail Designs of the Walker

Referring to Figure 2.3, we see the walker supporting itself from the tubesheet. In order to move about this planar environment, a pair of rotational joints are included between each of three vertical pins/mandrels. To move, the walker compresses one pin, retracts it from the tubesheet, moves in the plane with two degrees of freedom, and re-inserts the pin. When re-inserted, the pin is pressed against the tubesheet to account for any gradual loss of depth, and re-expanded. Subsequent moves are made by retracting the other pins, one at a time. The master control model for the walker comprises a series of magnetic (three-phase) servo transmitters connected to the joints of the walker. These provide feedback to the DC motor control of the four rotational axes. The controller also self-checks each connection from the master to the operating walker to ensure reliable control.

## 2.8 Conclusion

This chapter has described the overall design process via a FBD to indicate unambiguously the early stages of conceptual design decisions that were made in its realization. As noted in Appendix A, each of the several function blocks should also show the many allocations constraining the function. For example, *Segment Arm* would clearly show the 16 kg limit on units of equipment carried into containment. Literally hundreds of such constraints would be shown on a fully described FBD, laid open to scrutiny by the many engineers, managers, and operators involved in realization of the system. In this way a more traditional Functional Specification can be built with detailed conceptual knowledge of the *intent* of the designers.

As an aside, this two-part system was eventually replaced by a single, digitally controlled arm. Inspection of Figure 2.5 and Figure 2.6 will show that many of the functions remain unchanged even though a major shift in design took place. Many of the functional changes accrued due to the appearance of reliable personal computers and high-torque electric motors and drives. These functions existed "to the right" of the original FBD.

## Problems

2.1 Please locate the photographs/figures that describe the field environment and the field work in the nuclear service example provided.

2.2 What are halogens and why are they proscribed in a nuclear plant?

2.3 A walker suspended from a tube sheet as in Figure 2.3 could literally "walk out" of the tubes if there were a vertical error in a series of withdrawals and insertions needed for walking. How could this cumulative error be prevented?

2.4 If there were some lateral errors in positioning of a pin before insertion into a tube, how could step-wise walking be assured?

2.5 Given that the service arm has a reach of 1500 mm, what torque would be required at the "shoulder" (joint 2) to raise a 16 kg walker to the tubesheet?

2.6 Referring to Figure 2.16, a passive coupler component is in the center, the active receiver is on the left, and the receiving display is on the right (inner details in Figure 2.13). The passive component carries light sources and the receiver detects one of these. Please create an FBD for this portion of the two-part robot system and show where this FBD connects to Figure 2.6.

2.7 The model used in Figure 2.14 is half the size of the actual walker. Why is a 1-to-1 servo mode acceptable for driving this system?

## Notes

1. Westinghouse Electric (1979) US Patent 4,168,782, September 25, 1979.
2. Westinghouse Electric (1981) US Patent 4,295,740, October 20, 1981.
3. Westinghouse Electric (1982) US Patent 4,316,189, February 16, 1982.
4. http://asq.org/learn-about-quality/process-analysis-tools/overview/fmea.html.
5. Sturges, R.H. (1992) Reliability and safety in teleoperation, in *Safety, Reliability and Human Factors in Robotic Systems* (ed. J.H. Graham), Van Nostrand Reinhold, pp. 83–115.

# 3

# The Largest Mobile Robot in the World

We now describe another extreme example of mobile robotics, that of a large machine for hauling several tons of coal *per second* from the working face of an underground coal mine. This machine was designed to compete with the common practice of scooping up coal that has been mined (loosened from its seam) with several 4-wheel rubber-tired vehicles that are each driven in pre-selected patterns in the coal mine. The scooped-up coal is delivered by these vehicles by their drivers to semi-stationary conveyors that lead from the coal mine to the surface, there to be stored or transported by truck to waiting customers. Alternative mining approaches, such as "long-wall" systems are beyond our scope and involve very different kinds of automated processes.

## 3.1 Field Environment: Underground Mining

The field of underground coal mining presents a wholly distinct set of constraints from ordinary laboratory robot development. The interior of underground coal mines spans the gamut from "shirt sleeve" operations away from the working face of the mine to highly restricted areas with 100% humidity, temperatures over 100 °F, and potential ambient chemical levels that can cause harm to humans at daily operational exposures. Further, mine workers generally need to carry equipment through a maze of passages through uneven, dusty, and often muddy terrain. This terrain, as mentioned above, is also the environment of individual coal hauling vehicles in many installations. The floor of a coal mine is typically littered with fallen rock and trace amounts of coal, often muddied by the presence of water seeping into the excavated areas. Such a floor prevents accurate guidance and navigation by coal mining machines (dubbed "continuous miners") and haulage vehicles.

An underground coal mine often comprises a large area (several square miles) of carefully surveyed passages from which coal is first removed. As the coal is removed, the overlying

*Practical Field Robotics: A Systems Approach*, First Edition. Robert H. Sturges, Jr.

Companion Website: www.wiley.com/go/sturges

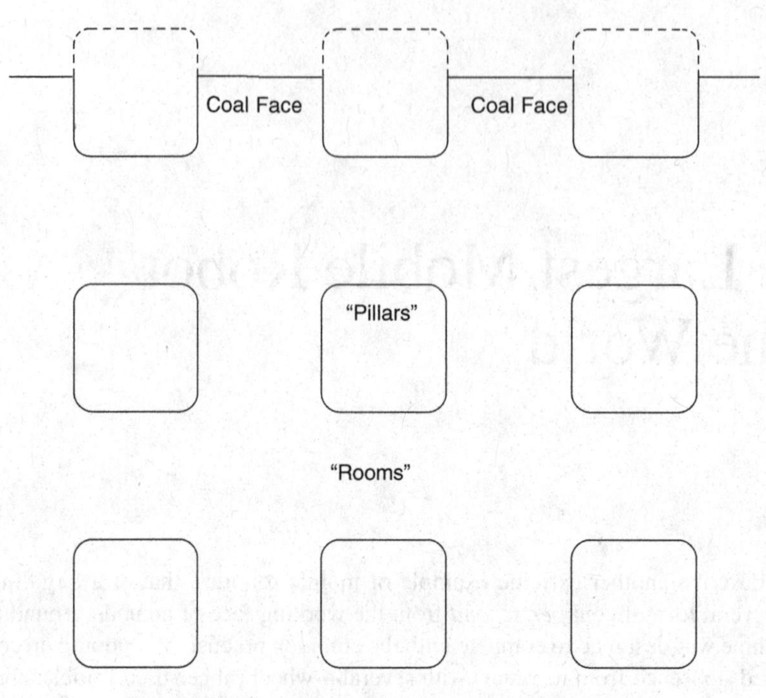


**Figure 3.1** An orthogonal room and pillar mine plan

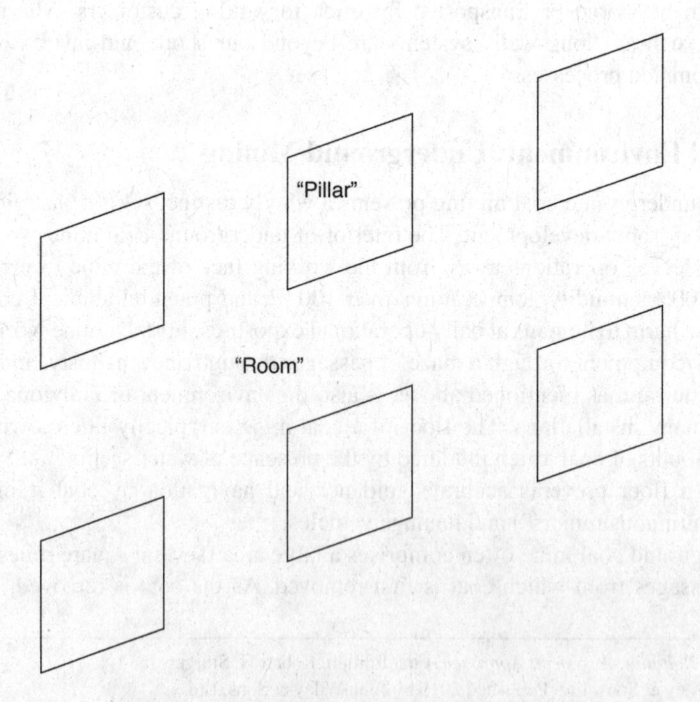


**Figure 3.2** An oblique room and pillar mine plan

rock becomes unsupported and must be reinforced. This process is the task of other manned equipment. The area, length, and width of unsupported "roof" created by mining the coal has been well-studied and standards are in place to ensure that "roof falls" are always remote possibilities. Such areas limit the access to large quantities of coal between scheduled "roof-bolting" tasks. The layout of a working mine comprises a set of "faces" that are to be gouged out according to a carefully laid-out plan. Mining the coal and removing it leaves behind a grid of pillars to support the remaining roof. Figure 3.1 shows a schematic plan view of a generalized underground coal mine that has been planned in orthogonal coordinates. Figure 3.2 shows an alternative design set at oblique angles. The coal not yet removed is shown in Figure 3.1 by dotted lines. In actuality, the area of the mine may extend several miles in each direction.

Infrequently, mining opens fissures that contain methane gas that must be ventilated out of the working area, so mines generally contain holes bored up to the surface for ventilating the mine. Sensitive detectors and moveable partitions are in constant use to prevent injury to miners from a concentrated gas stream. Another environmental condition involves lighting, which must be installed as the mining process proceeds, and this lighting must be made "intrinsically safe" so as to prevent the ignition of any methane from electric sparks or excessive, concentrated heat sources.

## 3.2 Field Work: Continuous Coal Haulage

The gamut of daily coal-mining activities are beyond the scope of this book, but we will address the major challenges of loosening the coal from its seam and carrying it out of the mine to the surface. As mentioned, there exist several methods to accomplish this task. In the past, railed carts were used to carry the coal, but these have been replaced by the introduction of several technological improvements. First, the coal is loosened from the seam by a machine called a "continuous miner." It comprises a rotating drum studded with hardened steel teeth, a set of caterpillar-like tracks for locomotion, and a driver skilled at cutting only the coal and not the floor and roof of the mine. The coal issues from the rear of the miner and must be continuously scooped up and removed. This task is carried out by transport vehicles that comprise an articulated body (for negotiating the sharp turns of a "room and pillar mine,") four large rubber tires, and a skilled driver that must follow a planned route from the miner to a delivery conveyor system installed at a central point and extending to the mine entrance and surface.

As an example, a continuous miner would be aided by a coal depth measurement sensor monitored by the driver adhering to strict safety rules (see Section 3.5) and a plan for coal removal. Such a plan considers the overall layout of the rooms and pillars of the mine, as well as the reach of the coal scoop vehicles. Since a continuous coal miner of the type considered here advances in small increments of "unsupported roof," the net effects on system performance are magnified with every movement taken. In any event the haulage must be carried out continuously along a series of connected "rooms." Pillars may remain in some plans, but in others the pillars are removed after the entire area has otherwise been cleared of coal.

Another popular method of mining coal involves the "long-wall" system illustrated in Figure 3.3. To begin, a series of pathways ("rooms") are mined out so that the extreme reach of the mine is cleared of coal *first.* Long rails (several hundred meters in length) that carry a shearing blade and removal conveyor are installed to mine the coal. The shearing blade is a flattened version of the continuous miner's studded drum. Also, a thin portion of the mined-out

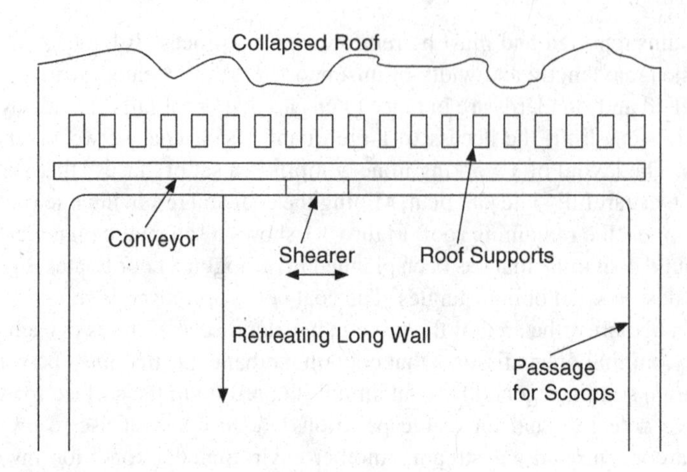


**Figure 3.3** A plan for long wall mining

area is fitted with a series of movable roof supports. The coal is sheared from the wall in a continuous process until the entire span has been completed. Then the entire set of machinery is moved back towards the mine entrance for a distance equal to the sheared thickness, and the process begins again. During this time of shearing, coal is constantly being delivered from the long wall conveyor to waiting coal scoop vehicles that carry the coal to a conveyor set in the mine. As the mining progresses, the conveyor is shortened periodically so that the scoop travel follows an optimal plan. Also during mining, the space left behind the movable roof supports remains unsupported and the roof regularly collapses according to plan.

In this type of installation, the only roofs requiring support are those that service the long-wall system, and no pillars are left behind. This method may induce large areas of surface subsidence, but the coal removed is almost 100% of the available area underground.

Generally, no guide-posts exist to assist in navigation, so the mining process in both types of installation is continuously surveyed.

## 3.3 Equipment Requirements

There have been many distinct types of equipment designed and utilized in underground coal mining, as outlined above. Here we will describe the design and construction of a very different type, namely a continuous miner driving one end of a series of connected conveyors that terminate adjacent to a relatively fixed central collecting conveyor. Such a concept eliminates the drivers and scoop vehicles that would ordinarily haul the coal from the face to a central conveyor (or two). The motivation for a new type of haulage system ("continuous haulage") is that the time needed to extract coal from a major seam (with heights of only 1 m and more) must be minimized per ton of product to achieve improved efficiencies. For this reason equipment costs and their maintenance must be balanced against the throughput of the entire system from continuous miner to the central conveyor. Figure 3.4 and Figure 3.5 show the function block diagram (FBD) of the beginning stages of the design and realization of such a system. It will be seen that the concept can be used for both "room and pillar" and "long wall" mining methods.

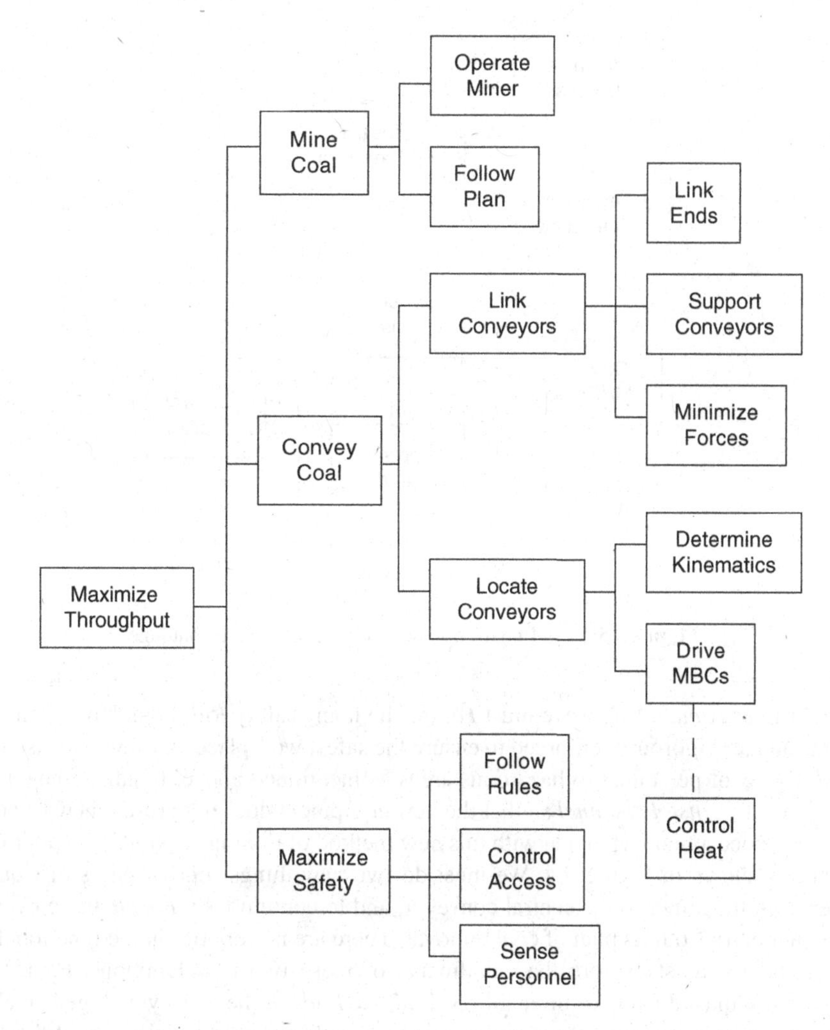


**Figure 3.4** Beginning of an FBD for continuous haulage. MBC, mobile bridge conveyor

We begin with a basic function: *Maximize Throughput,* wherein a perceived bottleneck resource is the manually driven coal scoop vehicles. The reasons for this perception should be clear: scooping is *not* continuous, vehicles need space and time to travel, and the need for drivers introduce personnel into the hazardous environment of the mine itself. The next level of functionality should also be clear, but it must be stated and agreed to by the entire team of designers, builders, and operators. It expresses that in order to *Maximize Throughput,* one must *Mine Coal, Convey Coal,* and *Maximize Safety.* To find the attending constraints on these functions, we create the next level of functionality and the many possible allocations that specifiy quantitative limitations. In order to *Mine Coal,* we need to do at least two things: *Operate* the continuous *Miner,* and *Follow* a removal *Plan* that comprises the mine layout, and the schedule for securing unsupported roof. We will see that the allocation for room and pillar sizes can be very different to that in the past.

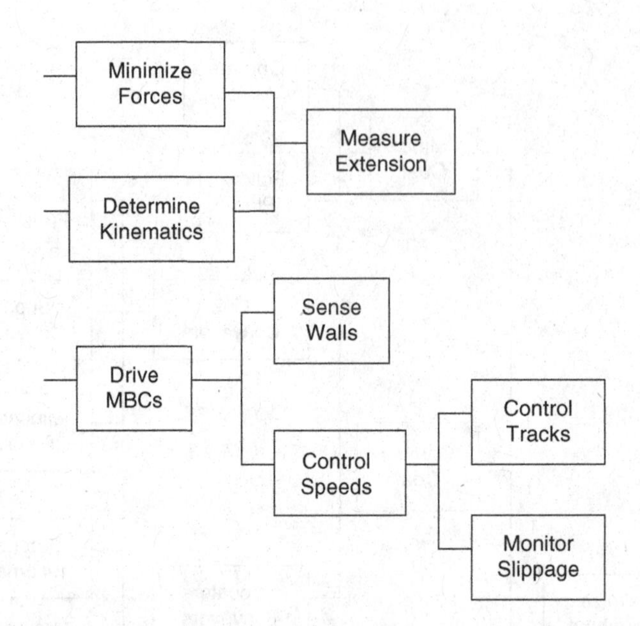


**Figure 3.5** Continuation of an FBD for continuous haulage

In order to *Maximize Safety,* we must *Follow* the many safety *Rules* that have been learned over time and are rigorously enforced to ensure the safest workplace possible. We also need to *Control Access* of personnel to hazardous areas as mentioned above. Finally, a new haulage system needs to *Sense Personnel* so that the new equipment does not present new hazards.

The large conceptual leap made with this new method of mining is expressed under *Convey Coal,* and is shown in Figure 3.4. We must do two new things: *Link* a series of *Conveyors* together from the miner to the central conveyor, and to continuously *Locate* these *Conveyors* as the miner carries out its plan of coal removal. There are no robotics in the functional *intent* thus far, and we must explore the entailments of these two new functions. First, to *Link Conveyors,* we must do at least three things: *Link* the *Ends* of the conveyors together, *Support* the *Conveyors* from the mine floor to make them movable and, in anticipation of the physics that will ensue, we must *Minimize Forces* between these conveyors.

Locating the conveyors is equivalent to locating their ends since they will be long, rigid structures. Alternatively, one may consider flexible conveyors as one coal equipment company had already done. Negotiating the narrow passages and tight turns in a room and pillar type of mine require us to *Determine Kinematics* in greater detail than simply operating the miner and driving scoop vehicles back and forth[1]. The method chosen for simplicity and potential reliability to *Locate Conveyors* was to *Drive MBCs*. Thus, each rigid conveyor structure in the chain is to be linked by a shorter drivable vehicle, as depicted in Figure 3.6. Each MBC (Mobile Bridge Conveyor) was propelled by steel caterpillar-like tracks and featured a short variable distance (lost motion) between its linked conveyors. Taken together, the cooperating MBCs and connecting conveyors formed the largest coal haulage system in use, spanning distances from the coal face to central conveyor of about 150 m. Each MBC needed to maneuver the pivots of the haulage conveyors such that turns were negotiated along a range of room and pillar layouts and in passages that vary in width. Ancillary equipment needed in the

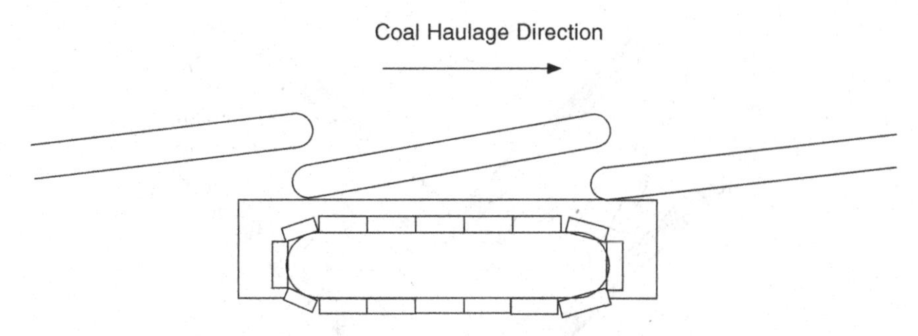


**Figure 3.6** Schematic elevation view of an MBC

mine surrounding the linked mobile system comprised lights and video cameras to obtain a coarse view of the operations. Drivers were connected by audio links to pass messages such as "slow down" or "drive more to the right."

In fact, a driverless system of this kind came *after* the successful demonstration of MBCs driven by mining personnel. An important side effect of having MBCs in use was the heat generated by their hydraulic drives. This fact gave rise to the functional consideration of *Control Heat* that affected the design of subsequent systems of this kind. Figure 3.4 shows the state of the system intent before robotics were introduced.

## 3.4 Conceptual and Operational Designs

An FBD of the next stages in the engineering of driverless MBCs and their sensors is given in Figure 3.5. At each branching of the diagram design decisions were made based on the known constraints and the "how–why" strategy of the FBD being built. For example, the pivots under the mating conveyors are provided by the MBCs which must automatically position the pivots consistent with the room and pillar spacing and the multitude of turns that take place every minute of operation, as shown in Figure 3.7.

Figure 3.5 shows that the functions of Linking and Locating are tied together at their subfunctions *Minimize Forces* and *Determine Kinematics.* These functions are met by a single concept: *Measure Extension.* This extension was included in the linking of driven MBCs and their conveyors. Knowing the extension and the angles between each linked conveyor completes a representation of configuration of the driverless system. One may well ask why extensions are even necessary, since one can conceive of a train running on fixed rails with no intervening extensions. This approach was found to be impractical, generating excessive forces between each linked section, both in driven and driverless modes of control. A little "slack" reduced lateral interlinked conveyor forces to near zero and gave additional flexibility to the process by which the MBCs were driven on uncertain terrain.

In order to drive an MBC, one could use a human driver that operated with very little knowledge of the units behind and before him/her. The audio links were in continuous use for critically important message-passing. Alternatively, Figure 3.5 shows a "robotic" approach that expanded into *Sense Walls* and *Control Speeds.* Finding the walls of the mine while driving through it proved to be a major step in driverless control since an algorithm operating with feedback could improve upon the abilities of a driver, especially under uncertain floor

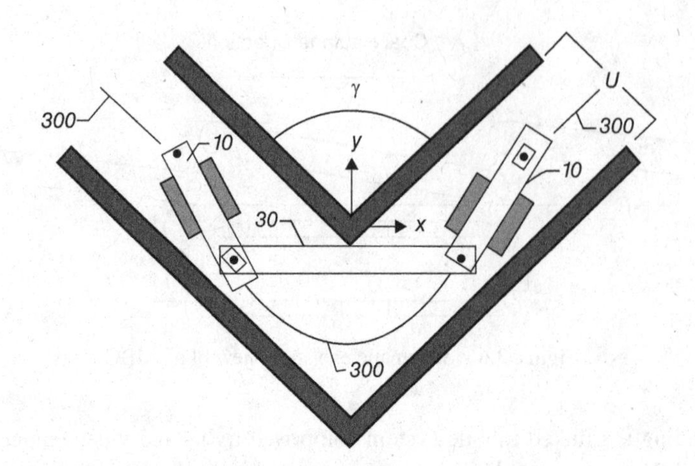


**Figure 3.7** Plan view of two MBCs and linked conveyor making a turn[1]

conditions. *Control Speeds* further expanded into *Control Tracks* and *Monitor Slippage*. Continuous measurement of the wall distance and orientation could determine the effects of slippage of the MBC's tracked vehicle design. These functions could not be performed by a driver.

## 3.5 Safety and Reliability

Many safety regulations in the coal mining industry have evolved over many years of experience, and these are beyond the scope of this book. It is the introduction of a new piece of equipment that gave rise to a new concern. A continuous haulage system would bring conveyors into the limited spaces of the room and pillar design, and these conveyors would be moving from side to side as they traversed straight and cornered sections of the mine. Such heavy equipment, moving tons of coal per second, would pose a hazard to personnel due to the changing limited spaces with nowhere to escape. The entire area in use by the continuous haulage system must then have strict access control.

For full-time reliability, the control module located on an MBC in a wet and dusty environment required self-checks and spare modules. The algorithm required forward and reverse operation, self detection, and driving around curves at any angle of the room and pillar layout consistent with the kinematics of the system.

## 3.6 Detail Conceptual Designs

Referring to Figure 3.4 and Figure 3.5, the function *Measure Extension* was realized by ruggedized potentiometers. Both extension and angle measurement were needed by the algorithm to help develop a graphical model of the entire continuous haulage system. A key device for *Sense Walls* was employed: a pair of laser-based distance-measuring devices that needed to be enclosed in an explosion-proof shell and still return measurements on the order of a few inches accuracy in a 360° swath around the MBC. With this feedback, a graphical model of the system was updated several times per second. The model served several purposes: it controlled the speeds of the driving tracks of the MBC while monitoring any slippage that might occur in the longitudinal and the lateral directions. Further, this model could determine if the driving space was indeed adequate for turning corners.

Also under *Control Tracks,* a previous hydraulic speed control system was replaced with an all-electric drive, eliminating the *Control Heat* side-effect. Although not mentioned in detail above, the discharging end of the continuous haulage system comprised a set of linear rails over the central conveyor. This design allowed the system to navigate tight spaces as the continuous miner moved from room to room at the coal face. In long-wall applications, the continuous miner was replaced by a head-end conveyor that needed only to move as the long-wall retreated.

## 3.7 Conclusion

A special-use mobile robot system and its description based on the Value Analysis/Value Engineering process has been presented. To date, it comprises the largest mobile robot in existence anywhere, with the ability to be expanded to even longer reaches. It serves also as an example of rigorous field robotic conditions in an industry that more typically employs manual methods. The advent of this new mobile robot system not only maximized throughput but also protects personnel from hazards in a high-productivity industry.

## Problems

3.1 When would "long wall" mining systems become impractical to use?
3.2 In light of the previous question, when would continuous haulage mining systems become impractical?
3.3 What type and how many sensors are needed on each MBC of a continuous haulage system?
3.4 Why is the trace of the conveyor pins (item 300 in Figure 3.7) a parabolic arc rather than a simple circle?
3.5 What factors contribute to the slippage of an MBC?

## Note

1. Virginia Tech and DBT America (2010) US Patent 7,076,346, September 24, 2010.

# 4

# A Mobile Robot for Mowing a Lawn

We now describe a more limited example of mobile robotics, that of concepting, designing, and operating a mobile robot for mowing a suburban lawn. We will examine the environment, tasks, and equipment as in the earlier examples, but will extend the function block diagram (FBD) to the point of identifying artifacts, including software. Alternative mowing approaches will be discussed, as well as suggested alternative functional pathways to success[1,2,3]. Our aim will be to enable the reader to duplicate the entire process of ideation through operation on a limited budget. One of the key limitations will be an allocation for total cost, which will be quite low.

As in the earlier examples, the iterative nature of design may lead us to put the design cart before the functional horse, but for the most part our discussion will follow the FBDs we create so that the "hows" and "whys" of every decision are clear.

## 4.1 Field Environment: Suburban Lawns

The field of suburban lawn-mowing presents another set of constraints distinct from ordinary laboratory robot development. We consider the practical conditions of a fairly level field (up to 5° from the horizontal) at its gross dimensions, and possible mole hills (lumps and dips in the 5 cm range). Wet weather operation may be required, and the field may contain an assortment of bushes, trees, and garden areas to be avoided by the robot's passage. Accuracy will be set by an *allocation* of ±5 cm. For example, Figure 4.1 shows a segment of lawn area that may reasonably be encountered.

Other field conditions include remote operation at a distance of several hundred meters from any control station, if the system is truly driverless and may not have an on-board computer with sophisticated sensors. (This condition may be waived if the mower was not remotely controlled.) Ambient lighting may vary from fully bright sunlight to operation in evening

*Practical Field Robotics: A Systems Approach*, First Edition. Robert H. Sturges, Jr.

Companion Website: www.wiley.com/go/sturges

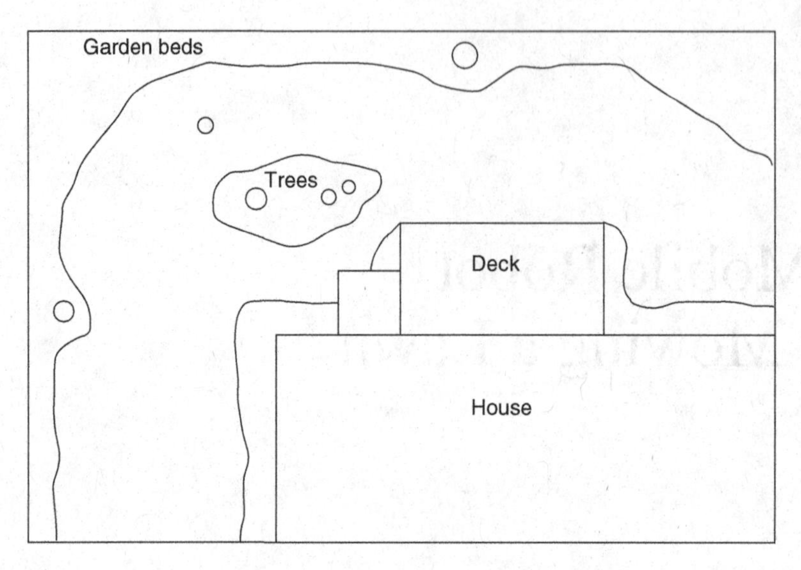


**Figure 4.1** Generic lawn scene with obstacles

darkness, with dappled lighting in between. Ambient sounds may include other nearby mowers (possibly gasoline-powered), barking dogs, songbirds, insects, and screeching bats. The sound level of the mower itself may be subject to local noise ordinances. Ambient electric fields (waves) may include local "wi-fi" areas, radio-controlled toys, and the ubiquitous cell phone. The presence of land animals and pets present a unique challenge that may require sensing and avoidance. Other ambient conditions include dust, sand, gravel, and rainfall.

## 4.2 Field Work: Navigation and Mowing

Daily gardening activities are beyond the scope of this work, but the basic function of *Mow Lawn* should be satisfied in theory and practice. The time needed to cover the entire cut area must be minimized to achieve a reasonable level of efficiency. For this reason, equipment costs and their maintenance must be balanced against the speed of the entire system from path planning to error recovery. Generally, no guide-posts exist to assist in navigation, so our intent will be to rely on the addition of such references in the environment. For example, Figure 4.2 shows an example of how one may *Plan Paths* and include sound receivers in key locations. The reasons for this selection will be made apparent shortly, and a method for finding these locations will also be discussed. Regular garden maintenance such as leaf raking or lawn litter removal is beyond our scope, but we will see that such activities could well be implemented with relatively minor functional changes.

## 4.3 Equipment Requirements

To be effective, a mowing robot must be able to carry enough energy to complete the mowing task without needing a recharge, thus the effective range becomes coupled with the paths planned for the machine. Also, the machine must be heavy enough to resist windage that is created or external, but not too heavy as to leave marks on the lawn as it moves or steers.

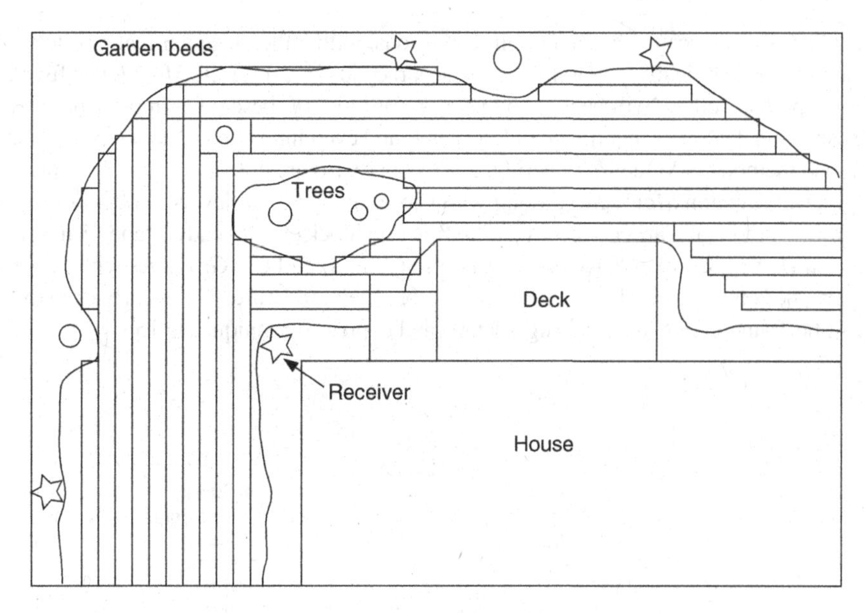


**Figure 4.2** Generic lawn scene with planned paths and added guide-posts

Moreover, the machine must be intelligently guided and avoid any random patterns that tend towards inefficiency.

We have not considered long-range mowing, such as needed for golf courses or commercial farm use. The function diagrams to be developed will show exactly where such considerations would change the design and operation for alternative purposes such as these.

## 4.4 Conceptual and Operational Designs

We initially examined six approaches to the problem of navigating a suburban lawn. They are all listed here for completeness:

Plan A: Fixed-mounted video cameras overlooking the entire lawn area[4].
Plan B: Robot-mounted video cameras seeking fixed objects for triangulation[5,6].
Plan C: Kinematic arms with joint encoders connecting a fixed point with the robot.
Plan D: Fixed-mounted sound receivers and robot-mounted sound senders.
Plan E: Fixed/robot laser scanners and robot/fixed light receivers.
Plan F: Radio wave distance measuring by amplitudes from fixed transmitters.

As an *initial* practical starting point for this case, we suggest a 20 m range and a robot speed of 1 m/s. We found that Plan A could not achieve reasonable accuracy of positioning at distances of 20 m and at reasonable speeds. It also could not operate with lawns that were even slightly sloped or filled with obscuring trees. Plan B worked well at 20 m, but required acquisition of three fixed objects, in this case striped poles, for triangulation. This approach required many such objects and suffered from additional ambiguity of positioning. Plan C was quite straightforward, but was limited to open fields without intervening bushes or trees. After initial field tests, Plan D became our technical favorite because everything worked as planned. Plan E could not operate in varying ambient light conditions, with either arrangement of lasers. Plan F was found to be technically impossible at the accuracy allocated for the *Mow Lawn* function. In the future, perhaps others will be able to make one or more of these other approaches feasible.

We can avoid the errors mentioned in Chapter 1, and even innovate in our designs by the well-established methods of Value Engineering. For example, Figure 4.3 shows such an expression of the high-level function *Mow Lawn,* along with successively higher levels of detail in a tree-like hierarchy. As before, the arcs connecting each function block can be read as "and." Thus, in order to *Mow Lawn* we need to do/have two things: *Plan Path(s)* and *Cut Grass.* As with the previous examples, these may seem obvious in hindsight, but they rarely are. These early decisions have the most profound effects on the design details and express our unique choices.

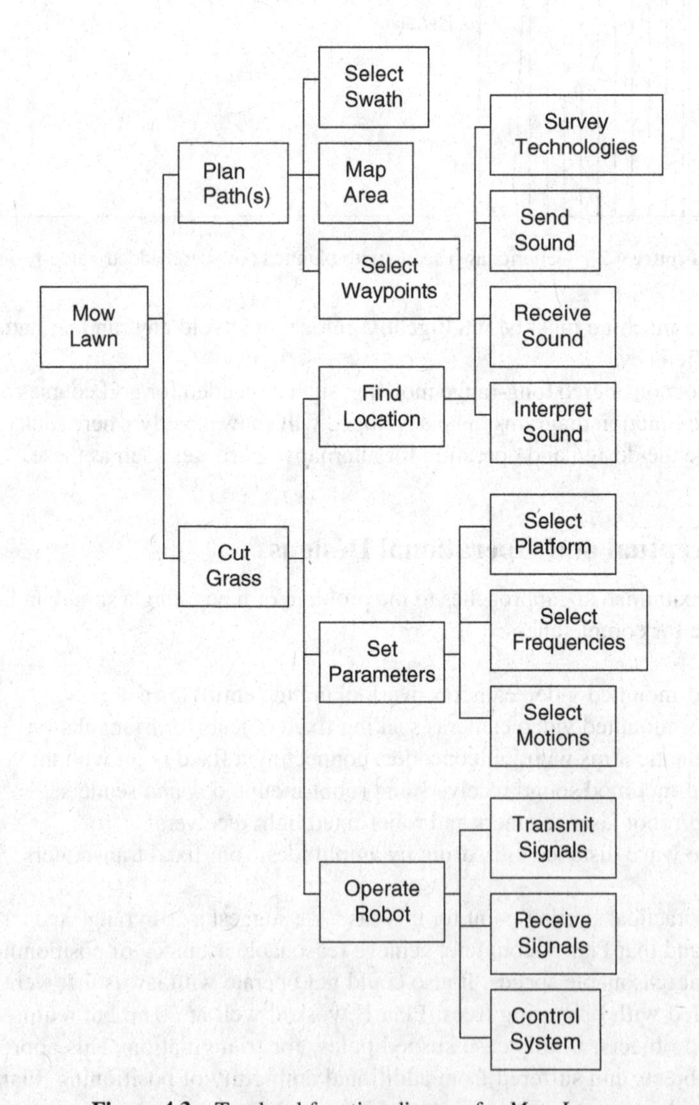


**Figure 4.3** Top-level function diagram for *Mow Lawn*

As an extreme example, we could have chosen to purchase a small herd of goats, for which there would be no path planning needed, and the grass would be eaten rather than cut. Obviously, other entailments would follow. More relevant to field robotics, however, is the very first decision to plan a path. Commercial "robotic" mowers do not have this ability (until now!), and instead mow an area randomly. Indoor "robotic" vacuum cleaners do the same thing. We put "robotic" in quotes to distinguish this enterprise of Field Robotics as the realization of a more intelligent automated process. We employ the term "practical" to mean that what we design and build must *work* and do so according to constraints.

## 4.5 Safety and Reliability

As mentioned in Section 4.3, the system must not cause harm when used as designed. Ideally, it should shut down automatically if control is found to be ambiguous or missing. Lawn mowers are well known to be dangerous machines when abused. We now CAUTION the reader that implementation of the system to be described will pose hazards if not monitored constantly. Several single-point failures have been identified, but not exhaustively, and no failure modes and effects analysis has been done. Therefore the descriptions to follow are to be taken for information ONLY and are not guaranteed to actually perform safely over any period of time. The instantiation of the system described depends highly upon the skill and care used in making the equipment and executing the given software. Since a wide range of readership is expected, we cannot be responsible for failures, unsafe conditions, or misuse of the ideas presented here. If these conditions are not accepted by our readership, one should read no further, nor excerpt sections of the foregoing discussions.

## 4.6 Detail Conceptual Designs

We should mention here that function logic, the means by which we reason our way from general to specific, is not normally encountered in the iterative cycle of "analysis, synthesis, and evaluation." It coexists with our "normal" way of thinking[7], and offers a great deal in terms of organization and expression of our designs to others. A detailed description of the process is given in Appendix A.

## 4.7 High-Level Decisions

At this starting level, we will outline the highest-level functions and their alternatives. Moving from left to right answers the question "how?," or "what is needed?" For example, in order to *Plan Path(s)* we need to do three things: *Select a Swath* to be cut, *Map* our *Area* to be cut or avoided, and importantly, *Select Points* along the way through which the robot should traverse. (Your function diagram may differ.)

First, the width and height selected will affect the map and afford us a means to quantitatively include lateral and vertical errors in positioning. The swath is different, because it may be wider (up to the width) or narrower in different places, and does not consider the grass cut height. We will set the height to a generic 5 cm, and the width to a modest 44 cm. (This value was determined by visiting a local lawn and garden shop to see what was most commonly available.) Next, to *Map* the *Area* involves many high-level decisions. The robot could be equipped to map the area by itself, recognizing obstacles and distinguishing lawn from paths. At this level of detail, we are free to explore and evaluate the options. For example, Figure 4.1 illustrates a generic lawn scene with grassy areas and obstacles detailed. Figure 4.2 shows this same scene, but with added bounds for the operation of the robot. Notice that Figure 4.2 includes locations for "Receivers." These will be discussed after more detailed functions are brought to light. (Design is an iterative process, even with function logic!) Once we have made these decisions, we finally consider selecting waypoints within those bounds. This decision can be contrasted with a method of lawn mowing that considers every point in the workspace of the robot. How we determine either of these points is another matter to be dealt with shortly.

To summarize, we have determined that three subfunctions are needed to express the way we have chosen to plan a path. Generality is still largely maintained, but will be reduced progressively in the discussions that follow. What we have already decided is not the only way to *Plan Path(s),* but the functional representation expresses our intent unambiguously.

To cut grass, we need to do three things: *Find* the *Location* of the robot at any instant, *Set* the scope of the technology by its *Parameters,* and finally *Operate* the *Robot* itself. We placed these three items in the vertical order shown since the function higher up on the page will affect, when qualified, those beneath it. In this case the decisions are sequential because they express logical constraints. Again, finding the location of the robot could be accomplished by many means, which we will explore. We refer to this activity as the "x, y problem" in robotics because the planar position of the robot is given by its Cartesian coordinates. Its orientation, or heading, is given by its rotation about the vertical axis (at least).

The rather generic *Set Parameters* function, or decision set, will begin to narrow our design options and set the strategy for solving many of the technical problems that we will encounter. If we neglect to do this task now, we will very likely spend a great deal of effort iterating the "analyze, design, evaluate" loop later. Finally *Operate Robot,* our ultimate demonstration, appears after much thoughtful discussion and dozens of decisions. Yet, the chosen approach to realization is still open to wide design latitude.

The *Cut Grass* function is also quantifiable with respect to a cost target at this point, so that all subsequent decisions are guided by a real practical value. This means the function also needs an "allocation" or quantity associated with it that is non-functional but constraining. We will set this value to be US$2000 (in 2014). Experts in marketing will no doubt have many opinions regarding this choice. We make it here to include as many readers as we can as practical aspirants to this practical field robotics example.

## 4.8 Conceptual Design—Technologies

We leave the highest level options behind as we make the next set of decisions. This is necessary to instantiate a concept that can be further detailed. Since we have left many options open, the issues of making or manufacturing can be addressed effectively. It is the aim of the field of "concurrent engineering" to do exactly this, so that unnecessary challenges are avoided in detail design and production of artifacts. We will now express the intent of our design concept functionally.

To find the location of our robot, many possible technologies exist. A short, and not exhaustive, list of these is shown in Table 4.1. We list the reasonable range and expected location error of each so that we can make an informed selection relative to function and cost. One of the popular tools of automatically guided land vehicles, the laser distance ranging device

**Table 4.1** Selected technologies for robot localization

| Technology | Range | Error |
|---|---|---|
| Dead reckoning (ODO) | Unlimited | Increasing |
| UT 40kHz, with radio sync | 10m one way | 3–5cm |
| UT 40kHz | 1m reflective | 5–10cm |
| IR Sharp™ | 1m reflective | 5–10mm |
| Tag line | 10m | 5–10mm |
| GPS | Unlimited | 30m |
| DGPS | Unlimited | 5–10cm |
| Buried cable | Unlimited | 10–20cm |
| Vision (CCTV) | 20m | 10–20cm |
| Compass | Heading only | 5° |
| Prepared field (magnets) | Waypoints | 5–10mm |
| Vision with processing | 20m | 5–10cm |

(LIDAR), is not considered practical for several reasons. First, it is expensive beyond our modest means, and secondly it is computationally expensive to employ. Thousands of research hours have been spent in universities all over the world making this technology work for their particular applications. Converting its outputs to the two numbers needed for the x, y problem is beyond the scope of this book.

In the same way, many of the technologies listed either do not match our need for accuracy (on the order of a few centimeters), do not operate to the range of the average lawn (a few tens of meters on a side), or do not fall within our cost constraints.

For example, GPS as found in your "smart" phone or on your dashboard represents brilliant engineering by many disciplines, including the launching of a dozen very expensive satellites. GPS-based solutions to the x, y problem work well at ranges of several tens of meters, but no finer. Its technical cousin, differential GPS (DGPS)[3], can resolve a few centimeters, but has practical limitations. It requires "line of sight" operation: nothing can be in its way, such as a bush or a tree. Also, it is too expensive in terms of our constraints.

Dead reckoning by odometry (counting fractional wheel rotations) typically operates without position feedback. It is subject to slippage errors in both distance and angle of turn, therefore we have considered its error as "increasing" with time or distance from a given starting point. Field ground conditions vary from place to place and may be predictable only to an experimentally determined range. A related approach using inertial guidance has become practical recently with the introduction of micro-electro-mechanical systems (MEMS). These devices suffer from "drift" in a different way than ground slippage, but also feature "increasing" error with time or distance. Occasional correction with a known reference would be needed, obviating an "unlimited" range.

Ultrasonic (UT) localization, also termed SODAR (sound detection and ranging), continues to be popular with hobbyists due to its low cost. Its weakness appears in the rapid decrease in range with increasing frequency, and the need to distinguish the desired reflections from significant "noise." Alternatively, a "one way" propagation of a sound wave in air would require a means to synchronize sender and receiver. We have demonstrated both light flashes and radio pulses to accomplish this. The error is not easily reduced since a stronger signal (at a lower frequency) would entail the discrimination of a less precise "wave front." Conversely, a higher frequency signal would attenuate much more rapidly with distance.

Distance ranging by infrared (IR) parallax measurement has been shown to be practical and inexpensive in, at least, the embodiments made by Sharp™. Exploring this promising area of technology beyond the present 1 m range could be fruitful, but has not be attempted here.

A tag line is simply a lightweight extendible cable tensioned with a motor or spring. Distance is measured by counting fractional pulley rotations. Error accrues from gravitational sag in the cable, and possibly snags along the ground in rough field surface conditions. Again, line of sight is required for distance measurements, and more than one tag line may be needed.

From its inception decades ago, civilian GPS has grown in popularity for automotive, nautical, and even airborne applications. These systems are based on differentiating the propagation of several satellite-borne radio waves, and are most familiarly limited to an error of about 30 m.

A simple buried cable, as used in electric fences for pets may easily proscribe boundaries for a field robot, but do not solve the x, y problem.

Robot localization in the field by "vision" has been and remains an active area of research. As cameras and computers both drop in cost, and algorithms improve, this technology appears most promising. Given the cost constraints of this case, a fixed focus vision system has been found (Plan A and Plan B) to feature a range of about 20 m with a resolution of 10–20 cm.

Geomagnetic sensing (electronic compass) readily gives robot heading. Combined with inertial guidance, the two may constitute a reasonable alternative to dead reckoning, with the caveats that its error increases with time and distance unless periodically corrected by some other means, and that nearby magnetic fields (from the robot motors) may cause additional inaccuracies.

Finally, one could prepare the field with magnetic materials spaced on a known grid. The "waypoints" thus created could assist in correcting dead reckoning technologies, but would stretch our definition of "field robotics."

By the process of elimination, at least, we are directed to choose a familiar technology: timing the flight of a sound wave. This approach suffers from systematic inaccuracies due to temperature and wind conditions, but lies within the practical range of our requirements. Entailments include synchronizing the sender and the receiver, since we cannot rely on reflective SODAR techniques in the field. There are simply too many echoes to distinguish, and the range in air is far too short. We will see that other technical hurdles will need to be leaped with direct send/receive sound methods, as we will address in the next section. Your choice of technology may be different, and several subsequent functions would need to be changed to reflect such a decision.

## 4.9 Conceptual Design—Set Parameters

In order to *Set* the *Parameters* for a practical field robot, we need to do (at least) three things: we need to *Select* the hardware and the software *Platforms* for the function of cut grass; we need to *Select* the practical *Frequencies* at which the sound system will operate; the frequencies of the implied radio link needed for synchronization; and we need to *Select* the kinds of *Motions* needed of the robot to accomplish its mission. There are many choices for each parameter to be set.

Considering the hardware, we should plan for a level of quality in the cutting of the grass which is consistent with the swath previously determined. The type of cut is a choice here, and is practically limited to rotary mowers, reel mowers, and string trimmers. More exotic approaches should be discussed to keep all options open and creative ideas "on the table." For example, one could consider a "hot wire" to burn the tips off silently, or a large-scale mower type reduced in size, namely a flail mower. Flail technology is widely in use on roadway berms to protect the equipment from rocks and stumps. Each decision will have entailments that need to be addressed as well.

Considering the software, we should plan on a level of skill shared by or desired by the design team. Choices here are broad, but have entailments in the software platform selection. For example, some languages are more prevalent in "embedded" designs, while others are found only in "desktop" environments. For example, BASIC (in one or more dialects) can be found on single in-line pin (SIP) devices costing a few dollars, while Excel™ (yes, it IS a language) has many quasi-real-time features that make it ideal for robotics. The language choice also may determine the vendor of the product, the integrated design environment (IDE) and the cost. A given assembly language usually appears on only one brand of processor chip, with its own proprietary IDE.

The issue of whether to select or avoid a proprietary language is a critical one. While most systems are readily available to a university, by license with a generous educational discount, some very attractive alternatives are marketed to industry with high prices and exclusion clauses for commercial applications. For example, LabView™ is widely known[8], used, and taught, but carries additional licensing fees if one would employ it in a commercial product. The language(s) to use comprise a long list (not exhaustive) of practically available choices and are listed in Table 4.2.

The convergence of hardware and software also comprise a choice, since the robot may carry its own computer, or be linked to one remotely. The rigors of field robotics entail a

**Table 4.2** Computer languages, platforms, and vendors[a]

| Computer language | Platform | Vendor |
|---|---|---|
| Labview | Windows | National Instruments |
| | Mac-compatible | National Instruments |
| BASIC | Stamps | Parallax |
| | Windows (VS) | Microsoft |
| | MindStorms | Lego |
| Excel | Windows | Microsoft |
| Open Office | Windows | Oracle |
| C | Mac-compatible | Apple |
| | Windows (VS) | Microsoft |
| | Windows (VSE) | Microsoft |
| C++ | Mac-compatible | Apple |
| | Windows (VS) | Microsoft |
| | Windows (VSE) | Microsoft |
| Python | Mac-compatible | Apple |
| | Windows (VS) | Microsoft |
| | Windows (VSE) | Microsoft |
| Java | Mac-compatible | Apple |
| | Windows (VS) | Microsoft |
| | Windows (VSE) | Microsoft |
| | Solaris | Sun |
| RDS | Windows | Microsoft |

[a] Trademarks are not listed here for brevity, but remain the property of their respective owners. We have not listed any varieties of the Linux platform, but would consider its use appropriate according to the skill and experience of the programmer.

ruggedized computer if robot-mounted, such as the Toughbook™. A little rainstorm could destroy an ordinary laptop or unprotected embedded system. We will find that this important function, *Isolate Weather,* will appear deeper in our function diagram. We will also see that mounting the principal processor away from the robot opens the prospects for relatively easy upgrades.

## 4.10 Conceptual Design—Operate Robot

In order to *Operate* the *Robot,* we will need to do many things. In Figure 4.3 we list the top-level three: transmitting and receiving signals (which we can consider as a unit) and the highly detailed *Control System.* The word "control" here is used as a verb, so it tells us to do something. What it will tell us his how to design the bulk of the system we need. It does not simply hold the place for an existing system.

The functions *Transmit Signals* and *Receive Signals* refer, of course, to the radio link we needed to establish so that our time-of-flight sound system could be synchronized. It also refers to the need to be able to control the robot from a remote location, whether or not there is a control computer on board. So we have a choice to make in this respect. In our case, we elected to remotely control the robot platform independently of the computer platform. The reasons for this choice should be made clear: computer technology advances so rapidly, that if we were to choose an on-board computer system, we may find that its integration with the other hardware may soon become a "legacy" problem. Separating the two empowers us to change the computer platform any time, either for upgrading or maintenance. This decision also affects the availability of the system positively, since a spare software system can be prepared and swapped into service at any time. Our experience with embedded systems also indicates that expansion of functions can be impeded with the "wrong choice" of processor and IDE, and much time and effort lost as a result.

In this regard, we should also mention the long history of teleoperation that preceded robotics as practiced today. By teleoperation, we mean operating a robot or other system remotely by wire or other link. Such systems were at first mechanical and employed to handle radioactive materials behind leaded glass. Today modern teleoperators assist in performing minimally invasive surgery, wherein the surgeon is located remotely from the patient. Such teleoperators have advantages when "computer-aided" techniques are used. For examples, minute tremors in the hands of a surgeon can be filtered out, and camera positioning can be made to automatically track the places of interest to the surgeon's remote eyes[9].

In principle one could perform all of the tasks in Figure 4.3 manually, and it would serve us well to include such an operating mode in case of a temporary failure of any part of the system. Moreover, the task of bringing the robot from its place in a garage or shed to the location where mowing is to begin may be better performed manually than pre-programmed. The grass-cutting may simply need to be varied from day to day. In fact, we will describe this operating mode and the equipment needed to employ it as a helpful feature for the robot we introduce into the field.

## Problems

4.1 Consider Plan A and Figure 4.1 together. How accurate could Plan A become when geometric distortion of the video image is taken into account? Please state your assumptions clearly.

4.2 Consider Plan B. How could triangulation work to get a definite location?

4.3 Consider Plan B. What ambiguities could arise with three fixed reference objects?

4.4 Consider Figure 4.1 and two alternative robotic mowers: a commercial "random pattern" mower, and the "planned pattern" mower of Plan D. How much longer in time would the random mower need to cover the same area as the planned mower, if they operated at the same forward speed?

4.5 Suppose a US$2000 allocation were made for the function *Cut Grass*. How would this decision affect each of its subfunctions, and the higher level *Mow Lawn*?

4.6 The Polaroid® SODAR transducer system operates by raising the gain (and adjusting the lens) according to the time elapsed between the pressing of the "take picture" button and the receipt of a reflected 55 kHz sound pulse. Determine a reasonable gain versus time plot. Be sure to state all your assumptions.

4.7 Referring to the previous question, why should the Polaroid system need to have any change of gain with time? Would it work for our application?

4.8 What is a "tag line" and how could it be used to *Mow Lawn*?

4.9 Suppose you wanted to operate this robot indoors. What are the entailments you would need to consider?

4.10 Consider another "exotic" way to *Mow Lawn*. What new technologies would be needed in FBD form?

4.11 Consider the three common methods to *Mow Lawn* described. Discuss the entailments of each.

## Notes

1. http://www.probotics.com/robotic-lawn-mowers/robot-mower-comparison, especially pages 2 and 3.
2. http://www.robotreviews.com/roomba/roomba-review-2.shtml (accessed June 30, 2014).
3. GPS Wing Reaches GPS III IBR Milestone, in InsideGNSS, November 10, 2008.
4. Kay, M.G. and Luo, R.C. (1993) Global Vision for the Control of Free-Ranging AGV Systems. Proceedings of Robotics and Automation, IEEE International Conference, May 2–6, 1993, Vol. 2, pp. 14–19.
5. Jia, Z., Balasuriya, A. and Challa, S. (2008) Vision based data fusion for autonomous vehicles target tracking using interacting multiple dynamic models. *Computer Vision and Image Understanding*, **109**(1), 1–21.
6. Isozaki, N., Chugo, D., Yokota, S. and Takase, K. (2011) Camera-Based AGV Navigation System for Indoor Environment with Occlusion Condition. Mechatronics and Automation (ICMA), IEEE International Conference, August 7–10, 2011, Kwansei Gakuin University, Sanda, Japan, pp. 778–783.
7. Ma, Y., Chen, G. and Thimm, G. (2008) Paradigm shift: unified and associative feature-based concurrent engineering and collaborative engineering. *Journal of Intelligent Manufacturing*, DOI 10.1007/s10845-008-0128-y.
8. http://www.ni.com/labview/; http://en.wikipedia.org/wiki/LabVIEW (accessed June 28, 2014).
9. Nishikawa, A., Taniguchi, K., Sekimoto, M. *et al.* (2010) Design and Control of a Compact Laparoscope Manipulator: A Biologically Inspired Approach, intechopen.com, pp. 365–380.

# 5

# The Next Levels of Functional Detail

## 5.1 Quantifying Conceptual Design

Figure 5.1 begins the practice of decomposing the higher-level functions into a few lower-level ones, and it should seem familiar. We have already discussed the three things we need to do in order to perform a reasonable survey of technology that we might best employ: *Find Accuracy, Find Range,* and *Estimate Cost.* We have set quantified limits to these foundational values. The function diagram of Figure 5.1 goes no further in detail, since the decisions have been discussed and made. We may consider them prescribed, as many function "allocations" are in business and commerce. For example, we may need to set a market price, which will determine a production cost that will probably determine the viability of an enterprise.

The appearance of the three subfunctions to *Survey Technology* on a separate figure are dictated in part by the medium in which we create, deliberate, and employ functional thinking with our design team. Indeed, the larger the physical area available, the more we can readily assess. We are limited here by the page size, but we will repeat the highest level (left-most) functions often to remind us of our place in the bigger picture. In practice the function diagrams appearing here were all initiated using small sticky notes, each bearing a single function. These were at first disorganized, but were easy to sort and form into a hierarchy as the functional needs of the project grew. Moreover, our intent began to take form with the literal repositioning of these functions with respect to each other.

Figure 5.2 shows one branch of the prior function diagram expanded by functional decomposition. It also shows that decomposing functions has a practical limit, as we recognize artifacts that satisfy the functional needs expressed to their immediate left. We will see that every subsystem and component in our robot is represented functionally. Every part must have a reason for existence, or else it is redundant and can be eliminated. An existing system can be reverse-engineered in this way to not only discover the details of its replication, but also to hint

*Practical Field Robotics: A Systems Approach*, First Edition. Robert H. Sturges, Jr.

Companion Website: www.wiley.com/go/sturges

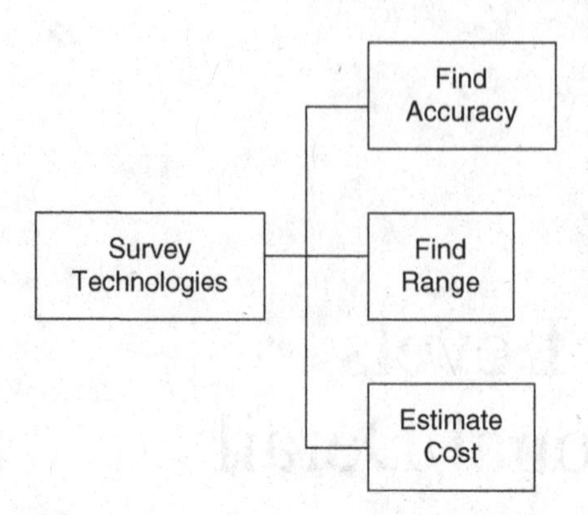


**Figure 5.1** *Survey Technologies* function diagram

at the intent of the designer(s). Further, by employing Value Engineering, we should be asking "what else will do the job?" as each function and artifact is selected.

## 5.2 Quantifying *Send Sound*

In detail, then, Figure 5.2 shows that we need to do at least three things in order to *Send Sound* for our time-of-flight distance measurements. At the hardware level, we need to *Generate* a *Sync*hronizing signal, *Transmit* it to the robot, and assure that there is equipment to *Receive* it. Each of these functions is subdivided once more into the detailed functions that can be satisfied by specific artifacts. First, we decide to send the sync signal as part of the pulse train sent by a radio transmitter, specifically NOT channels 1, 2, or 3, since they will be used to control the robot in its manual mode. At this point, we may appear to be "getting ahead of ourselves" since there is not much of an antecedent basis for this decision to *Set Channel 4*. It represents an example of the iteration that was performed in the design of this robot. At such junctures, one may simply suspend judgment, or read ahead to a related set of functions not yet described in detail.

Channel 4 was selected since it was otherwise not allocated, but we need to decide how to realize this choice. To the right of the *Set Channel 4* function are the answers: we need a custom-programmed chip to modulate the transmitter by computer control, and we need a way to get the control computer to send the signal itself. We will detail these artifacts when we have completed the discussion of all of the functions needed to *Send Sound*. At this high level, we have not specified that the sound sender should reside on the robot, or in the fixed infrastructure, since that choice depends on whether we place the computer on-board or not. In either case all of these artifacts will be needed, so our options for innovation and change remain open to us.

Since the logic chip running the *LMM3 Code* and the *Servo Board* need to be powered somehow, we explicitly decide on the (5 V) power of the USB line. This decision could well have been different if we decided to communicate with the Ethernet instead. A survey of available options showed us that there were far more USB-based functional elements that could connect to generic appliances than Ethernet-based devices. This survey informed us that the intent of the vast majority of Ethernet connecting boards are for communication rather than acting on the real world with programmable voltages and currents. A detailed look at how to

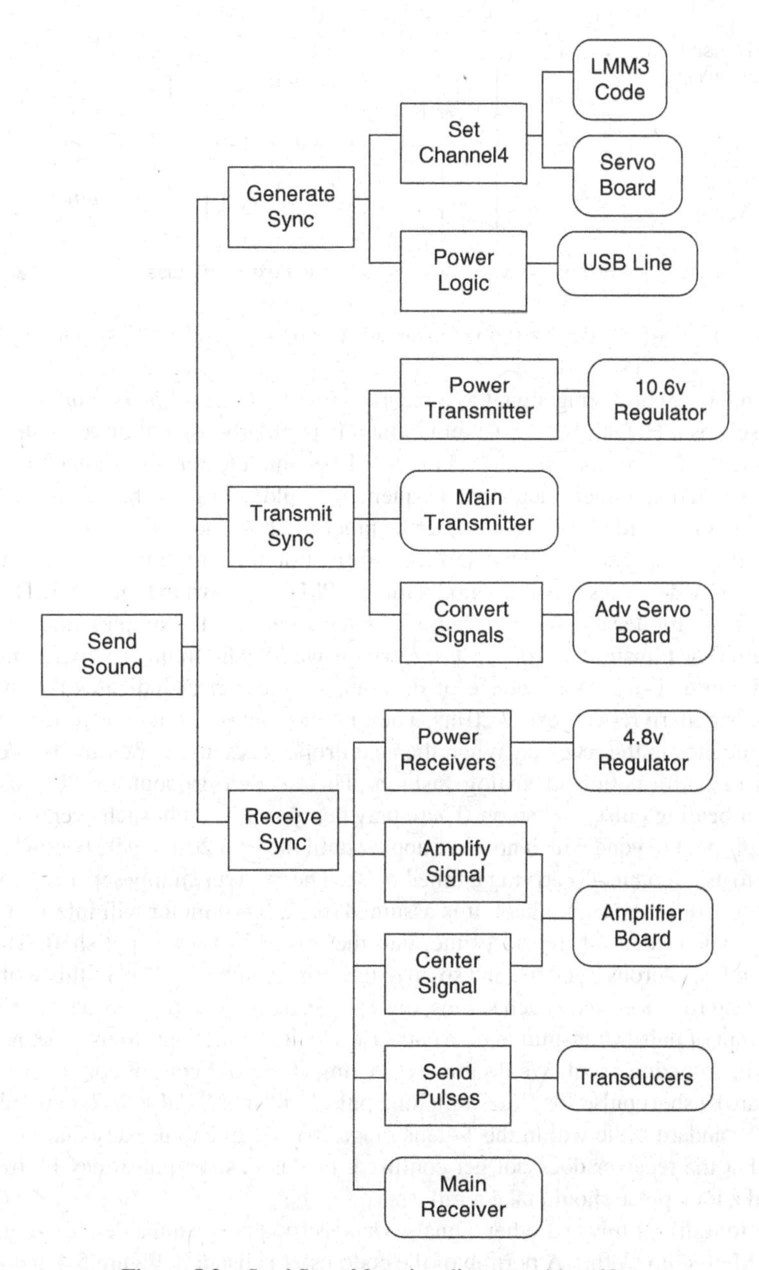


**Figure 5.2** *Send Sound* function diagram with artifacts

use the USB for such purposes is given in Appendix B. There you will also find code examples to match the hardware.

To *Transmit Sync* by radio link we need a transmitter, which we have termed *Main* in Figure 5.2 to distinguish it from another one which we anticipate using for manual teleoperation. Thus a major decision has been expressed: our control computer will communicate via a radio link in

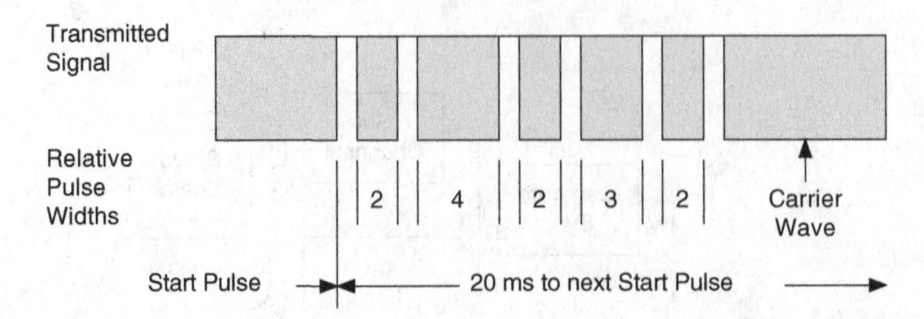


**Figure 5.3** Pulse position modulation (PPM) standard for R/C systems

order to send a synchronizing signal to a receiver. In detail, the *Main Transmitter* chosen for this purpose was a HiTech Laser 4™ unit, which is popularly used to drive model airplanes, boats, and cars. The reasons for this choice will become clearer as we quantify the *Select Frequencies* function. Other brands and frequencies could, of course, be substituted.

At this point it would be helpful to describe precisely how the *Main Transmitter* will work to transmit the sync signal. The standard for information transmission in radio control (R/C) equipment is based on pulse position modulation (PPM), as shown in Figure 5.3. This standard has had wide use for decades and continues to be the choice for the simplest and most reliable R/C medium. The transmitter broadcasts a carrier wave of radio frequency that is most simply modulated "on and off." A sequence of dropouts of the carrier indicates the information sent, and this pattern repeats every 20 ms. The first dropout (of about 300 μs duration) starts a timing sequence in the receiver. When the next dropout is sensed, the time between pulses represents the information in analog fashion. This second dropout also begins the next information-bearing pulse, and so on. There may be, typically, eight such events followed by a terminal dropout beyond which no more appear until the next 20 ms cycle begins. Each event is referred to as a "channel" carrying a timed pulse. The receiver simply separates each analog time event to a different signal lead. It is assumed that a servo motor will interpret that pulse width, now independent of the start time, into the position of an output shaft. These shafts control throttles, ailerons, rudders, and so on in the target equipment. (As of this writing, many soccer-playing robots employ such servos, driven in exactly the same way as described here.)

With a train of pulses transmitted on a carrier wave, it is our choice to use channel 4 to represent a synchronizing event. We do this by sending (from our control computer via a servo control board) a short pulse for "idle" or a long pulse for "sync." Pulse durations are specified by the R/C standard to lie within the 1–2 ms range. We are free to use any durations we like, provided that the receiver does not get confused. For us, a short pulse may be from 300 to 500 μs, and a long pulse should take a full 2 ms.

In order to transmit this and other signals, we specify a programmable chip (e.g., a PIC™ chip from Microchip Corp.). A portion of the code itself is listed in Figure 5.4 and is given in its entirety in Appendix C. In functional terms, this code simply receives the standard R/C pulse train sent via USB by our control computer, and modifies every fourth pulse according to Figure 5.5. Specifically, a sync signal (long pulse) is followed by three "idle" pulses. This sets the effective repetition rate for sending sound to 12.5 hits per second, or 80 ms between each event. This value was determined by the need to receive sound that may have taken 60 ms to reach the receiver, plus a margin for echo durations. The duration of 60 ms sets the upper bound of this time-of-flight system to 20 m. We will see later that this limit is practically

```
                ;LMM3.SRC       14 SEP 12
                ;This code translates PWM from the Phidget 1061
                ;into PPM that the transmitter needs
                ;adds a sync output
        DEVICE  PIC12C509,IRC_OSC,WDT_OFF,PROTECT_OFF,MCLR_OFF
        RESET   SETIO
OUT     EQU     GP0     ;Bitstring out
IN      EQU     GP1     ;Bitstring in
SYNC    EQU     GP2     ;Sync out
W1      EQU     11H     ;1st width in 10's of usec
W2      EQU     12H     ;2nd width in 10's of usec
W3      EQU     13H     ;3rd width in 10's of usec
W4      EQU     14H     ;4th width in 10's of usec
T       EQU     15H     ;timer byte
TIME    EQU     16H     ;saved time value

SETIO   CLR     GPIO    ;clean slate
        MOV     !OPTION,#11011111B      ;clear TOCS
        MOV     !GPIO,  #11111010B      ;two output bits

START   SETB    OUT     ;say hi by setting out to ready state
WFI1    JNB     IN,WFI1 ;idle while the input line is low
        CLR     W1      ;the 1st width time
INC1    INC     W1      ;start the count
        CALL    WAIT    ;wait 8 usec here
        JB      IN,INC1 ;if still high, continue counting
WFI2    JNB     IN,WFI2 ;idle while the input line is low
        CLR     W2      ;the 2nd width time
INC2    INC     W2      ;start the count
        CALL    WAIT    ;wait 8 usec here
        JB      IN,INC2 ;if still high, continue counting
WFI3    JNB     IN,WFI3 ;idle while the input line is low
        CLR     W3      ;the 3rd width time
INC3    INC     W3      ;start the count
        CALL    WAIT    ;wait 8 usec here
        JB      IN,INC3 ;if still high, continue counting
WFI4    JNB     IN,WFI4 ;idle while the input line is low
        CLR     W4      ;the 4th width time
INC4    INC     W4      ;start the count
        CALL    WAIT    ;wait 8 usec here
        JB      IN,INC4 ;if still high, continue counting
```

**Figure 5.4** A portion of code for translating pulses to the PPM standard

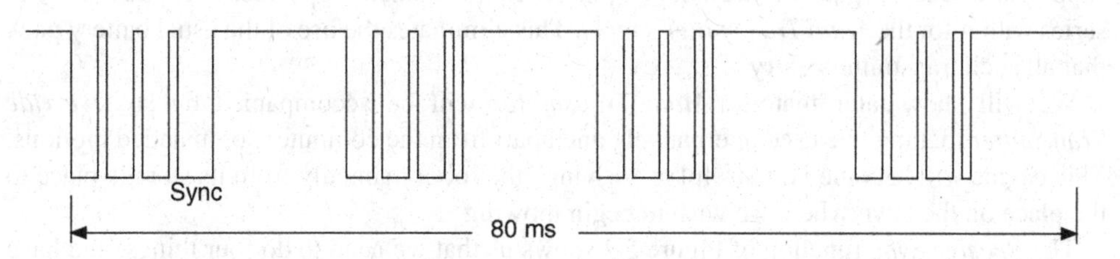


**Figure 5.5** Pulse train sent to the *Main Transmitter*

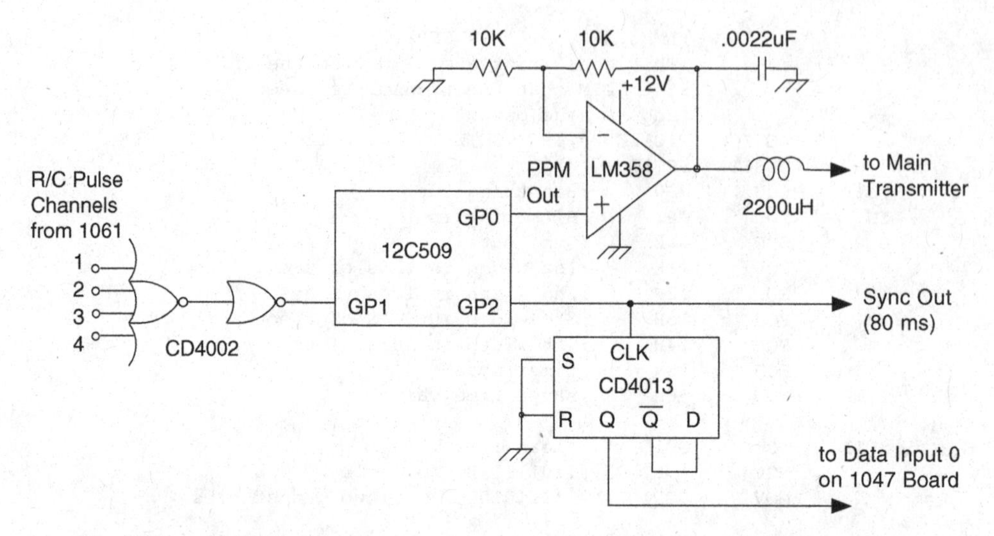


**Figure 5.6** Signal connections between servo board and *Main Transmitter*

achievable, and we will explain in detail why it is a real limitation on the distance measuring choice in Section 5.7.

The "servo board" mentioned is a device readily purchased from a number of vendors to translate the USB signals into standard R/C pulse trains. For example, we bought an SC-8000P from Tom's R/C, an Endurance R/C PCTx, and a Phidget #1061. The SC-8000P and the Endurance R/C PCTx are designed to translate USB signals directly to R/C transmitters via the "training" socket that most such transmitters have. The sync pulse would need to be generated from software for these devices. For more flexibility, we opted for the Phidget. The 1061 board features 8 channels of R/C pulses as sent by the control computer via USB directly to servos. Figure 5.6 shows the connections made from the 1061 device to the PIC chip and the *Main Transmitter.* Notice that an inductor has been included to isolate the radio frequency noise (the RF carrier) from our digital logic. Also, an LM358 op-amp has been included to raise the 5 V logical signal to the 10 V signal needed by the transmitter trainer input socket. The inductor serves to isolate the transmitter from our logic circuitry.

The logic chip does nothing to change the other channel information, just to periodically modify the channel 4 pulse width (position). Finally, we reverse-engineered the transmitter to find that the preferred supply voltage is 10.6 V, rather than the 12 V that is installed in our "USB translator" box shown in Figure 5.7. Many of the functions to be described are instantiated in this structure that directly connects via USB to the control computer. Since the 12 V supply is already regulated, we simply placed two avalanche-type rectifiers (1N4004) in series with it for the *Main Transmitter* supply. This eliminates the use of the usual battery pack that all such transmitters carry.

We will show later that the *Main Transmitter* will be accompanied by an *Override Transmitter* that may be used alternatively and apart from the computer commanded motions. This device will become very useful in "driving" the robot manually from its storage place to the place on the lawn where we wish to begin mowing.

The *Receive Sync* function of Figure 5.2 shows us that we need to do four things, and have in place an R/C receiver tuned to the *Main Transmitter* carrier frequency. At this point this

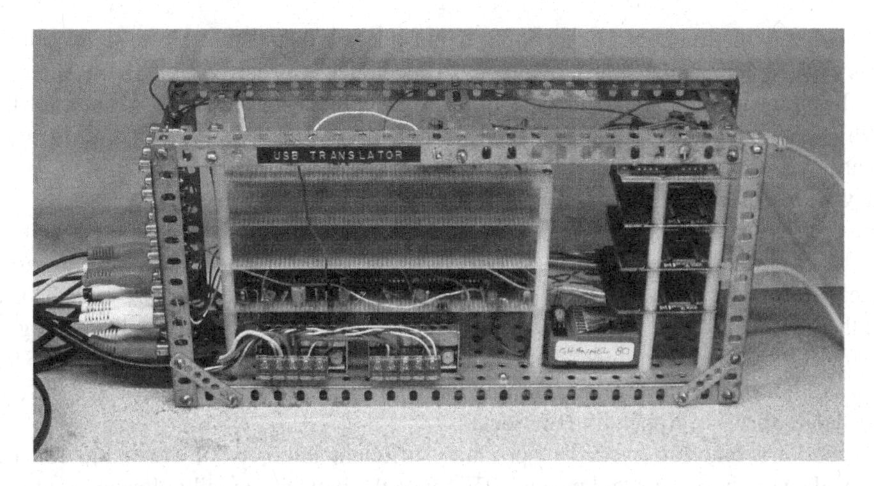


**Figure 5.7** USB translator unit needed to *Mow Lawn*

**Figure 5.8** Sound sending board for the field robot

frequency can remain unspecified. Indeed it can change at any time by either swapping a set of electronic crystals, or changing the transmitter and receiver devices to completely new technology. The key to flexibility in design is the R/C standard and the USB standards employed. The main receiver is usually a small plastic box with a wire lead antenna, intended to be mounted to a model airplane, boat, or car. It requires 4.8 V to operate properly, so this function is noted in Figure 5.2 and instantiated by an adjustable regulator chip, an LM317. The current drain depends on the servos driven by the receiver, and for our purposes is only a few milliamps.

The sound sending system comprises an amplifier board of our own design, shown in Figure 5.8. It must do two things: raise the voltage and current level of the sync signal while converting it to the frequency needed of the transducers and to "center" the signal about 0 V. In this case, 25 kHz is the target frequency. This choice was driven by two considerations. First, the attenuation of sounds in air doubles when comparing 40 with 25 kHz, according to ISO9613[1]. Secondly, the background noise (birds, insects, machinery) at 25 kHz is far lower than at lower audio frequencies. This latter observation was made by listening to ambient noise with a sensitive amplifier and a selective bandpass filter.

The duration of the 25 kHz pulse train is determined by the amplifier board and set to about 1 ms. This gives the transducers 40 repetitions at the target frequency. We will discuss this choice in detail a little later in Section 5.7. The "center signal" function represents the need to drive the transducers with a balanced "AC" signal, so that the maximum applied voltage does not exceed recommended limits. This signal is shown on the schematic of Figure 5.9. Not shown in Figure 5.9 is an LM78L05 three-terminal voltage regulator. Please see the recommended data sheets in Appendix D for details.

An ancillary function needed to produce sufficient intensity of the sound for a 20 m range is the supply of plus and minus 20 V. Since the batteries supply plus and minus 12 V, another 8 V is needed "on either side" of the batteries. This is accomplished most simply by inserting a pair of isolated DC-DC converters of a nominal 7 V each between each battery and the corresponding terminals of the LM675 power op-amp shown in Figure 5.9. We used a pair of unregulated 7 V units from Recom™ (type RN-1207S). Since the load is so low, about 12 mA, the received voltage is nearly 8 V, satisfying the 20 V requirement. Included on the amplifier board is a pair of 1000 μF filter capacitors, as well, but not shown in the figure. The LM675 op-amp dissipates less than 1 W in this application, but should have a small heat-sink attached to it to keep the junction temperature down on a hot summer's day.

Finally, the transducers themselves are selected, and there are many possible devices to choose from. We used Prowave™ 250ST-180 devices with good success. The number of transducers is important: each unit has a specified "beam width" that limits its signal output to a relatively narrow effective angle. (Please refer to Appendix D for details.) To maintain a fairly uniform wave front from the transducers, we mounted 12, wired in parallel, in a circle to an ordinary plastic (PVC) 2-in. pipe cap, as shown in Figure 5.10. We note that the distance between each transducer is small enough that no destructive interference between them has ever been a problem.

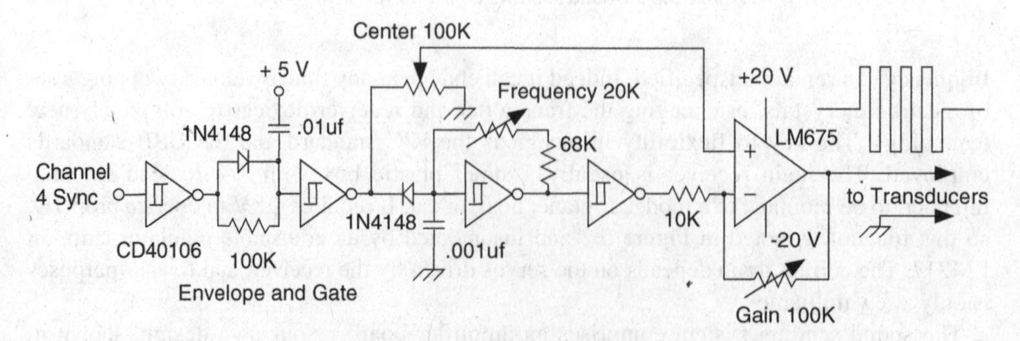


**Figure 5.9** Schematic of the sound sending board

**Figure 5.10** Sound sending transducers mounted for the field robot

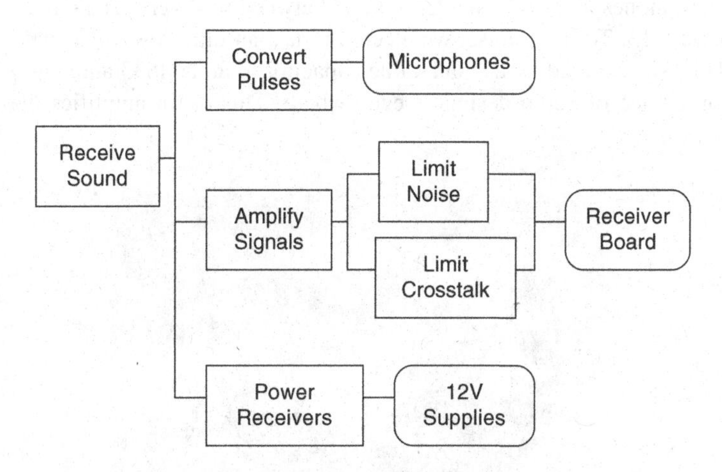


**Figure 5.11** *Receive Sound* function diagram

## 5.3 Quantifying *Receive Sound*

According to Figure 5.11, the function *Receive Sound* for the purposes of obtaining a time-of-flight are straightforward for those already skilled in analog/audio electronic design. For the general reader, we will describe in sufficient detail just what needs to be done and how to do it, so that one can specify its design to another. We must do three things: *Convert* the pressure waves sent by the transducers into reasonable electronic representations, *Amplify* those signals subject to a set of important conditions, and *Supply* the *Power* to do this. Again, it may seem at first obvious what needs to be done, but there are entailments that make even this set of functions challenging.

To convert pulses we choose a set of sensitive and selective microphones: sensitive in that we need to be able to distinguish the sent pulses from the ambient sounds, and selective so that we do not carry along the sounds of other frequency ranges. If we fail to do this, elaborate methods will be needed to sort out the signal from the noise. This we will avoid to keep the design simple, robust, and inexpensive to produce. The microphones selected were the Matsushita™ #0D24K2 and the Prowave™ 250SR-180, and both performed equally well in laboratory and field experiments. Many other types could suffice. Notice that the frequencies of the microphones do not exactly match those of the sending transducers. This choice is deliberate, since the response curves of these devices are complementary and usually spaced about 1 kHz apart. Details can be found in Appendix D.

Like the sending transducers, the number of microphones used at a single receiver is important due to the narrow angular range of reception. To determine the number, one needs to return to the map of the area to be mowed. If the receivers are positioned in fixed locations, rather than the senders, then the angular receptive range is found by sketching a "v" (actually a cone, in space) that encompasses the lawn portion within range. Our experiments have validated fairly uniform response when microphones are set 45° from each other, and in groups of either 4 or 6. Figure 5.12 shows one of the receivers packaged for use. We will now discuss the details of its construction.

The amplifier board we designed must do two things: it must limit the ambient noise picked up by the microphones and limit any "crosstalk" between receivers. The circuit schematic is given in Figure 5.13. To limit noise, we elected to use an ultra-low-noise audio op-amp, an LT1115C. This was coupled by a small value capacitor to an LM833 audio amplifier chip, a very common choice of audio designers everywhere. This chain amplifies the microphone

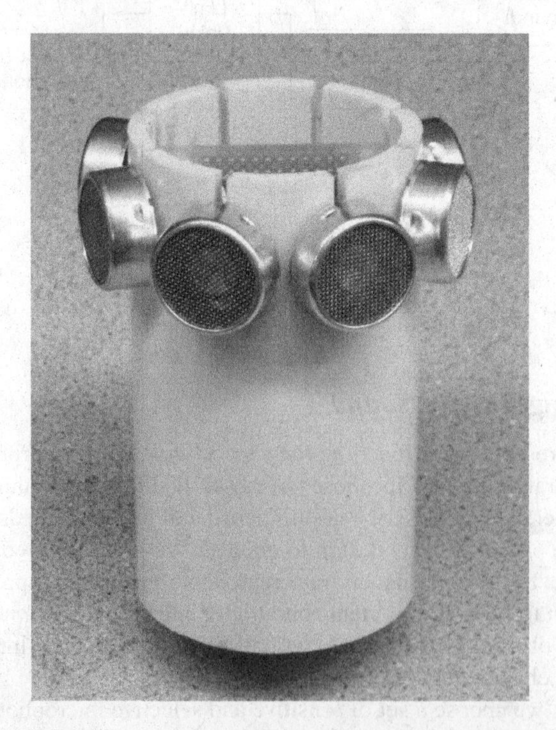

**Figure 5.12** A packaged receiver board with microphones

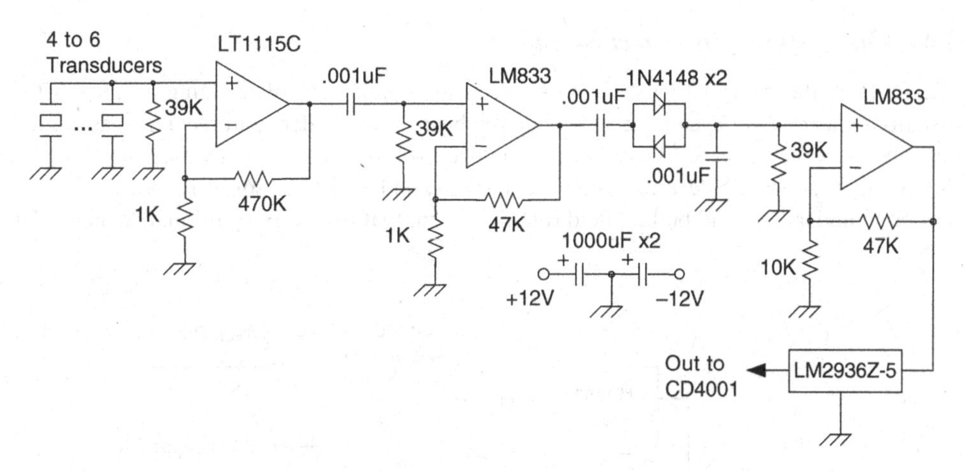


**Figure 5.13** Schematic of the receiver board(s)

signals by a factor of 22,000, or 87 dB. A gain of this magnitude in two stages prevents unwanted phase shifting and therefore instability. Next, the signals pass through a "noise gate" which cuts off any signal levels below about 0.7 V. This level was chosen since it was found through acoustic testing and measurement that birds, insects, and bats emit in this nominal 25 kHz range, and that their levels must not be mistaken for the pulses from the sender. Finally, another LM833 stage raises even the weakest signals to 5 V or more. This level is needed to drive the signal interpreter to be discussed next. We will stress that the choices of components is very broad, and that one can achieve our results with many other choices of devices. It is the functionality that we express here, so that the designer can be guided by relevant need and not just copy an existing design.

Crosstalk is the condition in which the amplified signal of one receiver interacts with that of another, since they share common 12 V power supplies. (Watch out! We just made another decision and did not express it functionally.) The reasoning behind this decision is that the infrastructure of field wiring is greatly reduced if signals can be shared in sets. We chose for our purposes the readily obtainable "category 5e" cable used for Ethernet applications over hundreds of feet apart. This cable features four pairs of wires, each wire of a pair twisted about its mate. This twisting is known to reduce crosstalk and susceptibility to external electrical noise sources, by essentially canceling out the unwanted magnetic fields. This cable is also very affordable.

In order to *Limit Crosstalk,* we elected to limit the current in each receiver board to a minimum, and not permit any current to be "spent" or dissipated in the system. To accomplish this, we inserted a low-dropout regulator (type LM2936Z-5) in the output signal line. It limits the output voltage to 5 V, and prevents any saturation of the digital receiving devices that it serves. In this way, the effects of even a very large sound signal in one receiver board has no effect on the sensitive amplifier stages of another receiver board.

Such details as above are critical in field robotics and not usually encountered on the laboratory workbench. The function *Limit Crosstalk,* for example, was not in fact anticipated as the receiver board was first designed. The "analyze, synthesize, evaluate" loop was employed when crosstalk was first noticed in a field test, and later eliminated. We will encounter many such anomalies as we investigate more functional details of this system.

## 5.4 Quantifying *Interpret Sound*

The function diagram of Figure 5.14 expresses the complexity of realizing a time-of-flight distance measurement system, given a fairly robust set of senders and receivers, as we have described in general earlier. Here are over a dozen subfunctions with a variety of artifacts at the far right: some are devices, some are systems, and some are software. All of these are needed to understand and build a field robotic system that *works.* In addition, take note of the

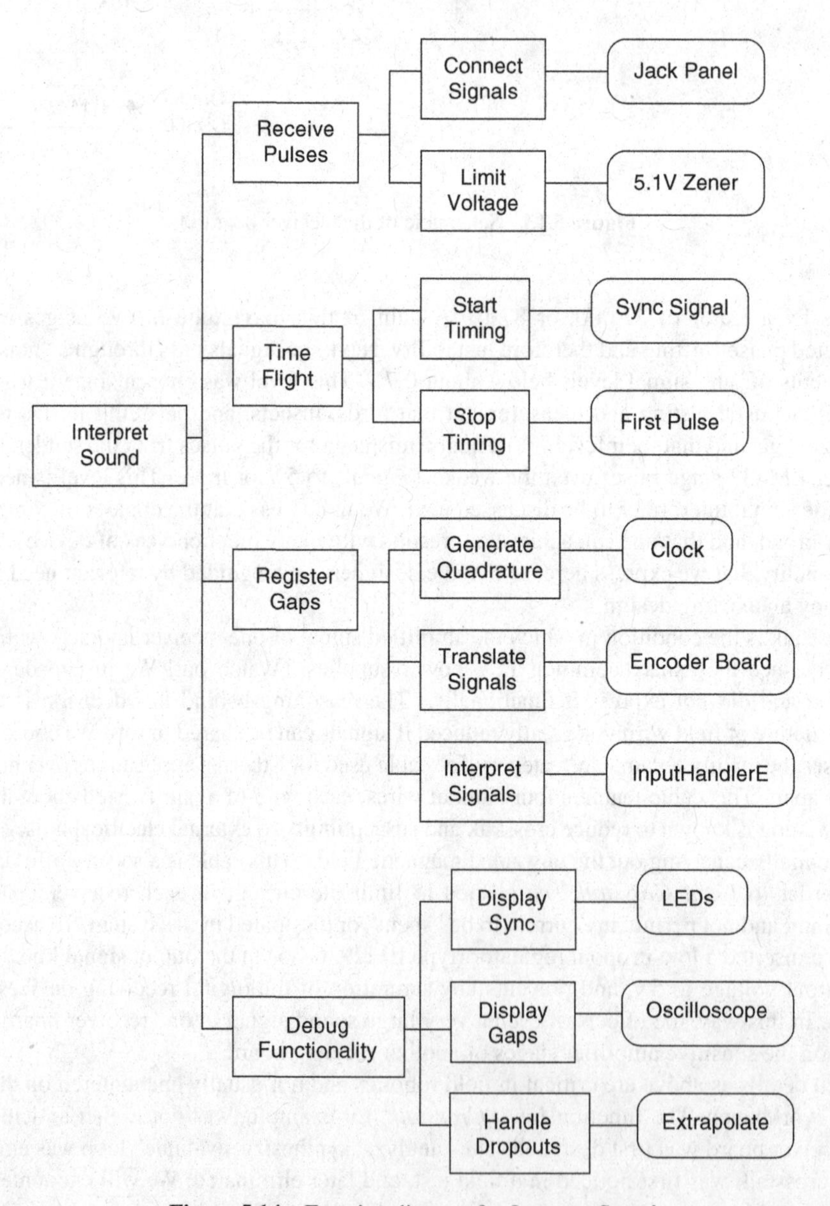


**Figure 5.14** Function diagram for *Interpret Sound*

fourth subfunction under *Interpret Sound: Debug Functionality.* Its existence acknowledges that we will make mistakes in judgment or execution of an idea. It accounts for the errors we may make along the way in realizing our system and displays for us the information we need to diagnose problems. It also takes into account those conditions under which the system may lack the feedback it needs to proceed automatically.

First we notice that to *Interpret Sound,* we need to receive the electrical pulses from the receiver boards mentioned earlier, and to connect them to our "USB translator" box. This latter device is not explicitly called out by name anywhere in our function diagram simply because it performs several functions. A single name tells us very little of what it is to do and how it will do it. Looking on one end of the translator box shows one physical realization of connect signals: a panel of RCA jacks. These were selected because they have low noise potential, are readily available, and are inexpensive. Recall that the signals being received here are analog, not digital. This implies sensitivity to noise pick-up externally, and we need to avoid that. In converting analog to digital, we need to satisfy the constraints imposed by the digital technology, namely the allowable voltage ranges, and the expected currents. This important function is provided by the inclusion of a 5.1 V Zener diode in the input section. The complete electrical schematic of this portion of the USB translator box is given in Figure 5.15. We will be referring to it with respect to the functionality to come (Figure 5.14).

To time the flight of a sound pulse, we need only do two things: start and stop a timer. We start the timer with every sync pulse. These were generated by the *LMM3 Code* discussed in Figure 5.4. Here is where we use them. To stop the timer, we "listen" for the arrival of the first electrical signal representing the incoming sound wave. The schematic shows an embodiment of these functions using a simple set/reset flip flop. Again, there may be many ways to accomplish the same functions, and the methodology we have applied leaves us open to new ideas and innovation.

You may have already guessed that reading the timer value will be necessary to finding the time-of-flight. We embody this concept by functions that describe in detail how this is to be done, rather than merely name an artifact called, say, "counter." In order to interpret sound, we need to digitally register the gaps between the sync and the first pulse events. To do this we searched for counter devices that also have USB connectivity so that our control computer can receive and use these values. Alternatively, we could have designed one, but it would need to do and have several of the same things. We selected an encoder board #1047 from the Phidgets™ Company, because it has several built-in registers that are driven by external timing pulses. While its expected application is reading the pulse trains produced by shaft

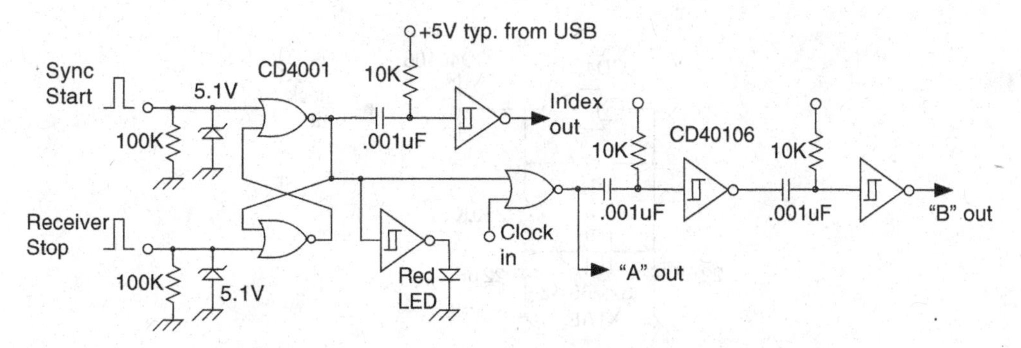


**Figure 5.15** Schematic portion of the *Interpret Sound* function

encoders, it performs the functions we need here in almost exactly the same way, namely translate signals from the timing gap we sense into computer-readable digital words.

Most incremental encoders are designed for forward and reverse motion, and so their counters are configured to count "up" as well as "down" to register shaft positions. In order to sense direction, the concept of *quadrature* was long ago introduced. Quadrature is illustrated in Figure 5.16. It comprises two input pulse trains, A and B, that are spaced 90° apart in time. Sensing which of the two pulse trains rises before the other determines the direction of shaft travel. The #1047 board requires two such input pulse trains, A and B, for the same purpose. So far we have only generated one, thus the new function *Generate Quadrature* is needed. This function is accomplished by the timing clock of the circuit in Figure 5.17, along with a few other components that simply delay the single clock pulses by a few microseconds in order to produce a synthetic quadrature signal. We will never need to reverse the order of these pulses, but the #1047 board requires both sets anyway. A conversation with the technical support staff of Phidgets confirmed this need, and a simple bench test showed our synthetic "channel B" worked as expected (Figure 5.15).

Finally, to instruct our control computer to interpret the *Gaps* in time between the beginning and the end of pulse train A, a software interface with the #1047 is needed. This code segment is shown in Figure 5.18. Before we detail its every line, it is important to recognize that the computer program that operates the robot may not be in sync with the output of the *LMM3 Code* on the PIC chip of Figure 5.4. One also does not know when a received sound pulse

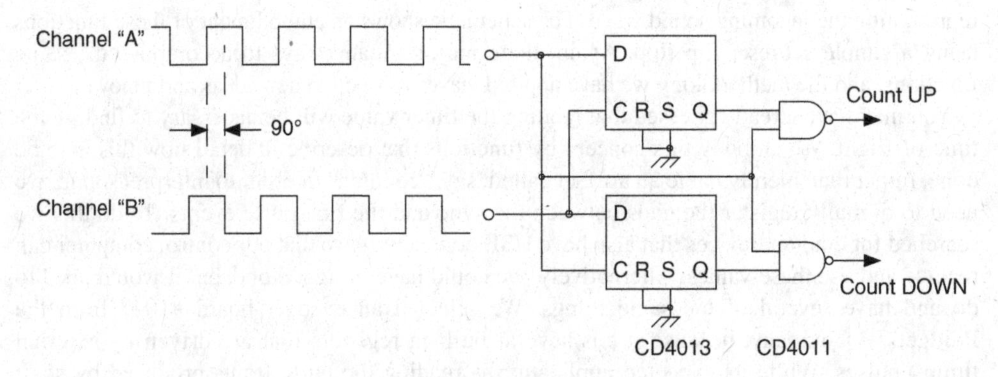


**Figure 5.16** Details of a quadrature signal for driving encoders

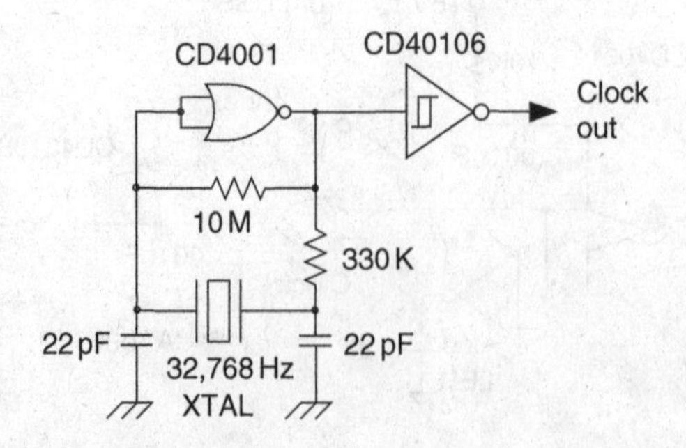


**Figure 5.17** Schematic portion of the clock circuit for *Interpret Sound*

```
111  int __stdcall InputHandlerE(CPhidgetEncoderHandle ENC, void *userptr, \
112                              int Index, int State)
113  {    //This is an encoder board event-triggered call-back function on sync
114  int iposition;                          //get only 2 positions, ca & cb
115  CPhidgetEncoder_getIndexPosition(ENC,ca,&iposition);//get the true
116  val[ca]=iposition-old[ca];              //position based on the index, which we set
117  if(val[ca]==0) val[ca]=old[ca];                   //tied to USB-translator hardware
118  else old[ca]=(float)iposition;                    //every 80 ms. Save the prior value
119  dist[ca]=val[ca]*100/SOS-OFFSET;                  //convert to inches
120  CPhidgetEncoder_getIndexPosition(ENC,cb,&iposition);//get the true
121  val[cb]=iposition-old[cb];              //position based on the index, which we set
122  if(val[cb]==0) val[cb]=old[cb];                   //check out section && notes
123  else old[cb]=(float)iposition;                    //every 80 ms. Save the prior value
124  dist[cb]=val[cb]*100/SOS-OFFSET;                  //convert to inches
125  if(dist[ca]<0 || dist[ca]>900) {                  //check range
126       Beep(200,10); manual=0;                           //if out, warn & set flag
127       flag[ca]=1; dist[ca]=smoothed_dist[0][ca];}  //extrapolate dist[ca]
128  else {flag[ca]=0; manual=1;}                           //reset flag
129  find_line(ca); avg[ca]=(float)bi;                 //pass line through history
130  da=avg[ca]; push(ca); pushflag(ca);               //post the avg distance da
131  if(dist[cb]<0 || dist[cb]>900) {
132       Beep(200,10); manual=0;                      //sounds like a click
133       flag[cb]=1; dist[cb]=smoothed_dist[0][cb];}  //extrapolate dist[cb]
134  else {flag[cb]=0; manual=1;}                           //reset flag
135  find_line(cb); avg[cb]=(float)bi;                 //and save the predictions
136  db= avg[cb]; push(cb); pushflag(cb);              //post the avg distance db
137  update_display2(avg);                             //   show off distance data
138  return 0;
139  }
```

**Figure 5.18** Program code segment for InputHandlerE

will occur, since it will depend on the varying distance between the sender and the receiver. For this reason, the code is instantiated as a standard "interrupt," that is, the computer will take notice of the appearance of the sync pulse whenever it happens to occur. To fully understand what is happening here, take notice of the connection between the PIC chip and the #1047 board. The PIC chip drives a flip-flop that changes state every time the sync pulse is generated. The #1047 board receives this pulse train through a digital port that recognizes a change of state. It is this change of state event that causes a USB signal to be sent to the control computer and cause the interrupt. The interrupt, in turn, causes program control to temporarily switch to the code segment of Figure 5.18. When the instructions have been completed, control returns to the main program where it left off.

One other feature of the #1047 board needs to be explained. Its counters (there are actually four of them) are reset by an "index" pulse that is produced by the "home" position of many incremental rotary encoders. The 1047 board is not compatible with the alternative "absolute" encoder. This resetting action takes place with every sync pulse of our system, since we will have connected that encoder input line to our sync pulse generator, as shown in Figure 5.9. The result of this action is that the encoder board will begin counting at the start of every sync pulse, and stop counting as soon as a sound pulse signal is received. The counter then waits for the computer code to initiate a query which sends the number along through the USB medium to the calling code.

If the above process seems mysterious, please review it again and follow along with the software in the interrupt module `InputHandlerE`, which we will now describe in detail. The code segment of Figure 5.18 begins with the type `int __stdcall` to indicate to the C-compiler that the following function actually is to be called only on an interrupt basis. (By the way, in C++, this function would be called a "method." Since C is a subset of C++, we can use a C++ compiler without changes to the code itself.) The name `InputHandlerE` was created by us in the `open_encoders` function that will be discussed in detail later. This latter code instructs the compiler to call `InputHandlerE` every time the #1047 board detects the sync signal. The timing of these events is best illustrated in Figure 5.19. If we consider the sync pulse to start every interpret signal event, three things will happen. First, our flip-flop on the translator board will start timing the gap, and signals "A" and "B" will begin loading a register in the #1047 board that begins to count "up." Also, the change of state sensed by the #1047 will trigger the execution of the code by interrupt. Finally, the counted value already present in the register

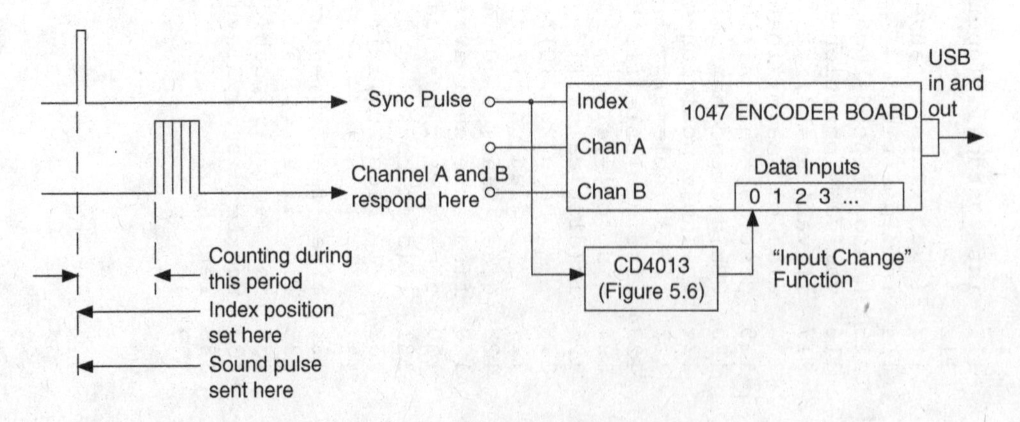


**Figure 5.19** Timing of the *Register Gaps* function

will be sent via USB to our control computer. After a brief interval (<80 ms), the gap ends and the "A" and "B" signals stop. The value accumulated in the #1047's register will remain at that value until retrieved by the interrupt call and reset by the next sync pulse.

We have not functionally decomposed each C function, but have adopted the practice of commenting every line for the reader. The code does, after all, follow a set flow chart pattern. One will also notice that we have simplified the code by never using the (dreaded) `goto` statement. Here is how the code works, line by line.

We declare an integer `iposition` to hold the value from the encoder board register. The leading "i" stands for the count since the previous index signal was sensed rather than the total accumulated count in the register. We then query the 1047 board for that value. The ca in the calling sequence is a subscript that keeps track of which of the four counters on the board we mean. For illustration purposes, let `ca==0`, and consider only the first ("zeroth") of the encoder counters. Next, we subtract the previous value accumulated and save the result in `val[0]`. We test for a zero value, which indicates that there is no count since the last call (perhaps a data dropout). If no data, we use the previous reading. Otherwise, we save the most recent reading in `val[0]`. The distance between the sender and receiver represented by the count is computed based on the speed of sound in air, `SOS`, and is offset by a value that accounts for the delay in detecting the fourth channel pulse (the long pulse) by the amplifier board.

Since at least two distances are needed for triangulation of robot position, the code repeats with a different subscript, `cb`, which for illustration could be simply a "1." We do not use a "for loop" for only two values, since the code will execute faster without that overhead. With two distance values in place, we then check to be sure that it is not negative or greater than a value that would represent more than 80 ms of gap time. Either case would constitute an error. If an error occurs, then the computed distance is extrapolated from the previous distances recorded by the function `smoothed_dist()`. This code will be detailed later in Section 7.2. An extrapolated value in fact rarely ever occurs in all of the field tests we have made of the robot. The `Beep` in the procedure can be heard every time one starts up the main code, since no values have been yet recorded. The short `Beep()` does not interfere with the timing of the procedure, it actually sounds like a click, and indicates that the startup of the system is normal.

The distance value reported to the locating functions, to be discussed at length later, is not the value just tested, but a slightly modified value based on the function `find_line()`. This function returns this modified value in the variable `bi` which is then saved in the `avg[0]` place. The purpose of this extra function is to intelligently smooth out the distance values based on recent history. There may be some jitter in the values received, and we have measured this jitter to actually be about 2 or 3 cm. To remove this jitter, `find_line()` constructs a plot of the previous 10 distance values along an x-axis, and using an RMS technique, passes a best-fit-line through these points. (See also Section 7.2.) The values nearest the y-axis of the plot are the most recent in time as shown in Figure 5.20. The slope of the line represents the trend in changes to the distance values over the past 800 ms. The y-intercept is the predicted value the distance should be if there were no jitter, and is returned in `bi`. This name recalls the straight line formula `y=m x+b`, where `b` is the y-intercept.

The code then repeats with the other subscript. Finally, the distances are displayed on the user screen, and the interrupt terminates. The foregoing discussion shows exactly how to interpret the signals. The code could be changed, of course, to do the same thing in other ways. In particular, additional receivers could be checked to find all of the current values. The geometry of the field we suggest would require only two such tests at a time, however.

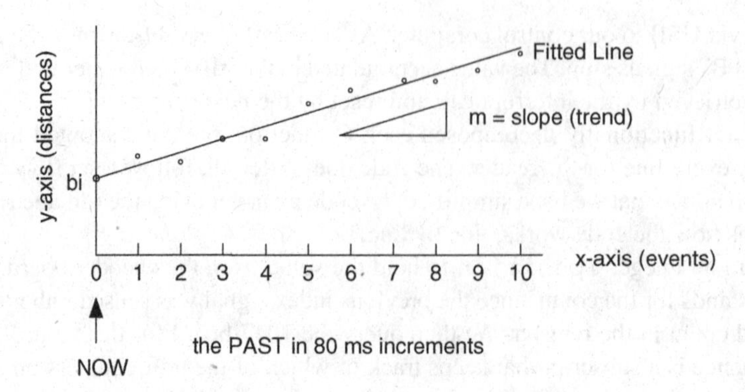


**Figure 5.20** The find_line() function as a diagram

The last subfunction needed to *Interpret Sound* (in Figure 5.14) is to *Debug Functionality.* Here we recognize the need to be sure that all of the prior system activities are working as intended, and we do three things: *Display Sync, Display Gaps,* and *Handle Dropouts.* A single LED in the translator box is driven by a separate radio receiver tuned to the same frequency as the *Main Transmitter.* By seeing this LED flash 12.5 times per second, steadily, we are assured that the robot is also receiving the sync signals and sending sound as expected. This LED is "off" otherwise. Each of the set/reset flip-flops on the translator board also drives an LED during the "off" state when it is not counting clock pulses for the encoder board. (Please see Figure 5.15 for details.) Thus, long distances appear as a series of short flashes, short distances as nearly continuous brightness, and no light at all for lack of received pulses. In normal operation, these LEDs flash at the same rate as the sync LED, and indicate any reception problems for each of the receivers in the field. Random noise will be seen as irregular flashes.

To actually "see" the sync and time-of-flight gaps, we temporarily attach the probes of an oscilloscope to the sync and set/reset inputs of the translator board. In this way Figure 5.21 was created. It shows what one may expect from the system with two receivers operating if the *Send Sound* and *Receive Sound* functions are working as designed. It also shows on the bottom trace the output to the *Main Transmitter* with all of the pulse-position-modulated "channels" in operation.

Finally, the *Handle Dropouts* subfunction refers to the broader condition of losing distance measurement data for extended periods of time (several 80 ms cycles consecutively). This situation occurs when the robot moves behind a tree or bush and when the alignment of receivers and robot prevent locating by triangulation. Figure 5.22a shows a generic triangulation problem with measurement errors. It is easy to show that even small distance errors can create a large location error. Figure 5.22b shows a possible alignment of receivers and robot in which triangulation is not possible within the ranges of error in distance we otherwise accept as normal, that is, a few centimeters. The `extrapolate()` software function was written to address this and the second problem inherent in triangulation: the ambiguity of position caused by not having enough data to distinguish which of two triangles to use. One could design the system to always receive and interpret data from three sources, but we have not chosen this option. We limit the number of receivers needed for any "patch" of the lawn to be mowed to just two.

As a brief aside, robot location by a vision system mounted to the robot has been long explored in the laboratory. Recognition of objects in the field of view of the robot always needs to include three points of reference in order to triangulate with measured angles rather than measured distances. We will eschew this option as overly complex and unsuitable for field application. In the future, improved processor technology and additional laboratory research may make this a viable option.

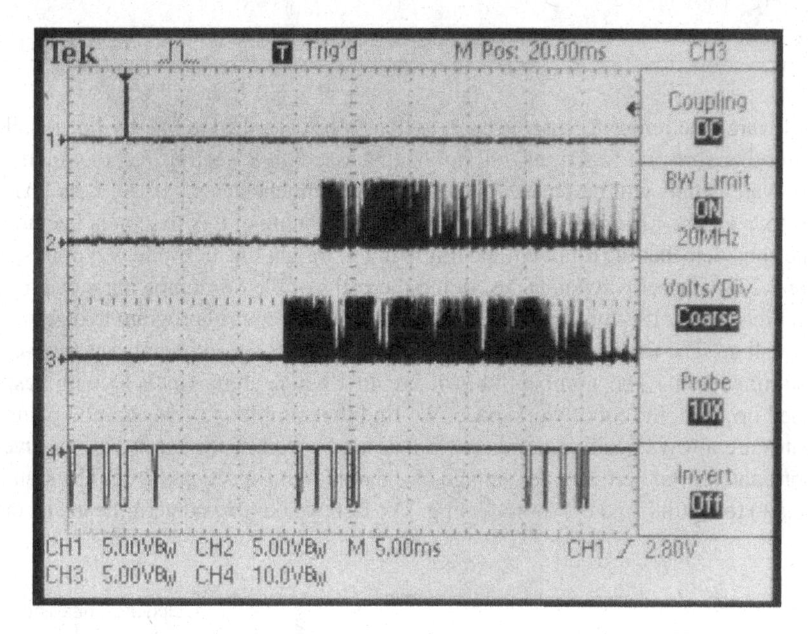


**Figure 5.21** Oscilloscope traces for *Send* and *Receive Sound* functions

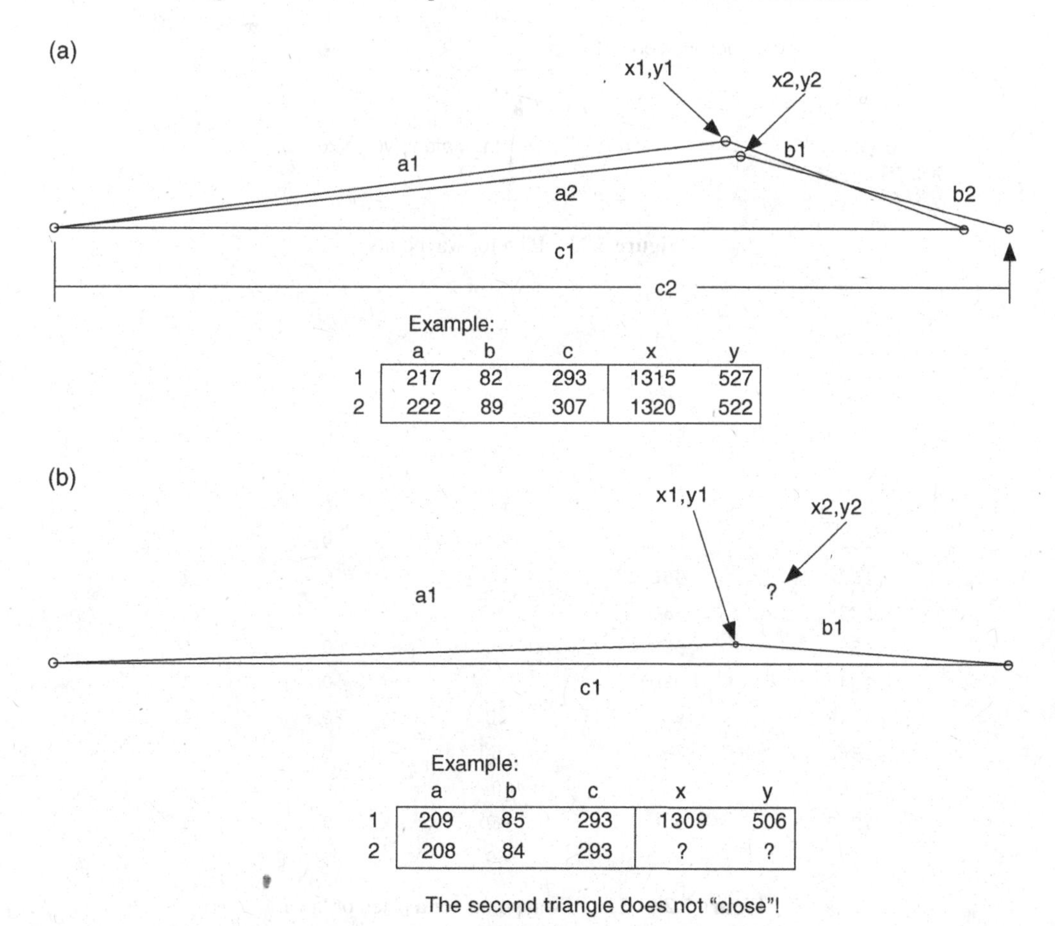


| | a | b | c | x | y |
|---|---|---|---|---|---|
| 1 | 217 | 82 | 293 | 1315 | 527 |
| 2 | 222 | 89 | 307 | 1320 | 522 |

| | a | b | c | x | y |
|---|---|---|---|---|---|
| 1 | 209 | 85 | 293 | 1309 | 506 |
| 2 | 208 | 84 | 293 | ? | ? |

**Figure 5.22** (a) Geometry of a triangulation problem due to error; and (b) geometry of an unsolvable triangulation problem due to error

The software function `extrapolate()` will now be described in line-by-line detail. Its connection to other software functions will not be discussed until Section 7.2, in which the entire set of software details will be given. As the leading comment states, "`use ten historic points to find the next in line.`" The meaning of this statement becomes clearer when one considers that the robot will be following a straight line between waypoints, as shown in Figure 5.23. A list of waypoints describing the paths of the robot mowing a patch of lawn is shown in Figure 5.24. It is this type of file that is read by the control program to determine where the robot will go. The first number is the *x*-coordinate and the second number is the *y*-coordinate of the starting point. The numbers shown are in English units (inches) with respect to a preselected origin in the robot workspace. We find that simpler patterns can be planned if the coordinates are always positive numbers. The desired speed of the robot will be given by the third coordinate, given here as a percentage of the maximum speed possible. The sign indicates direction: (+) for cutting and (−) for retreating. The last two numbers determine the receiver iden-

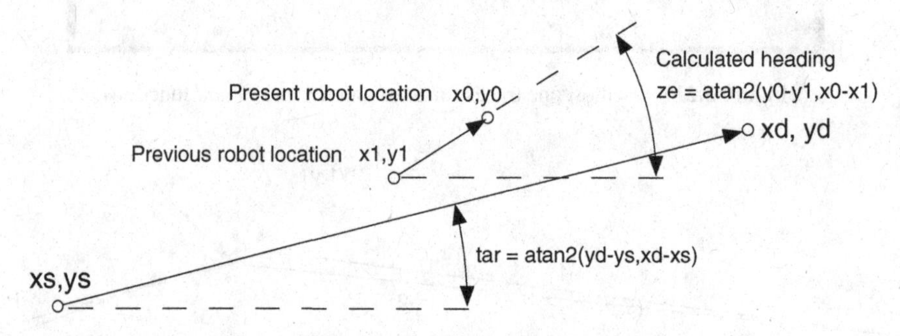


**Figure 5.23** Plan for waypoints

| xs | ys | vel | r1 | r2 |
|---|---|---|---|---|
| 1300 | 380 | 60 | 0 | 1 |
| 1300 | 950 | 60 | 0 | 1 |
| 1312 | 380 | -60 | 0 | 1 |
| 1312 | 950 | 60 | 0 | 1 |
| 1324 | 380 | -60 | 0 | 1 |
| 1324 | 950 | 60 | 0 | 1 |
| 1336 | 380 | -60 | 0 | 1 |
| 1336 | 950 | 60 | 0 | 1 |
| 1348 | 380 | -60 | 0 | 1 |
| 1348 | 950 | 60 | 0 | 1 |
| 1360 | 380 | -60 | 0 | 1 |
| 1360 | 950 | 60 | 0 | 1 |
| 1372 | 380 | -60 | 0 | 1 |
| 1372 | 950 | 60 | 0 | 1 |

**Figure 5.24** List of waypoints for a patch of lawn

```
423   int extrapolate() {                //use ten historic
       points to find the next in line
424        float delx=0, dely=0;
      //for the ideal delta for x & y
425        delx=(float)(cos(tar)*abs(vel/10.));
      //estimate from tar
426        dely=(float)(sin(tar)*abs(vel/10.));
      //estimate from tar
427        xe=delx+stack[0][0];
      //add it to the last x-value
428        ye=dely+stack[0][1];
      //add it to the last y-value
429        printw("Extrap'd xe=%d, ye=%d. ",
      (int)xe,(int)ye);
430        printw("delx=%5.2f, dely=%5.2f. ",delx,dely);
431        return 0;
432   }
```

**Figure 5.25** The `extrapolate()` function in C-code

tities that will be used to triangulate. Their positions must be known precisely by careful measurement from a fixed reference point. The process of creating this file could in principle be automated by giving just the corners of the "patch."

Given that the robot will be following a preselected straight line makes `extrapolate()` simple as shown in Figure 5.25. Changes in x and y (`delx` and `dely`) are declared as floats since integer arithmetic will be too coarse for our needs. The line that the robot will follow has a computable heading angle in the plane, in this case called `tar` for target angle. The expected changes in x and y if the distance data were available are computed from this geometry and the commanded speed of the robot. Scaling the percentage velocity by 0.1 approximates the actual distance covered by our robot during an 80 ms cycle. These expected values are added to the previous locations for x and y stored on a stack that is many rows in length. These previous locations may have been computed from distances, or from earlier calls to `extrapolate()`. Thus, the more cycles extrapolated, the less accurate the estimates will be. This should be expected since the heading commanded and the actual heading may differ. For the operator, these x and y estimates are posted to the computer screen. We tend to ignore them, but an extended number of extrapolated values (say, more than 10) indicates that something went wrong. Future software may address this unlikely condition.

## 5.5 Design Choices—Setting Parameters

Returning to Figure 4.3, the next function to be considered is *Set Parameters*. This function enumerates several high-level allocations needed to quantify the design of our robot. Specifically, we identify the need to *Select* a *Platform* for both the hardware and the software, *Select Frequencies* of operation so that equipment can be evaluated, and *Select Motions* so that robot geometry can be set. We will examine each in turn.

## 5.6 *Select* a *Platform*

Figure 5.26 shows the four essential "how" questions that need to be answered in order to *Select* a *Platform*. First, we need to select a type of mower. As mentioned earlier, we have at least three choices: rotary blade, reel type, and string trimmer. Others may be added to this list for your own design purposes. For several reasons, we selected a reel type. First, the quality of cut is the best of the three. The grass is not shredded at its tips, so it is less likely to turn brown later. Support of a reel on two sides naturally averages the smaller dips and bumps in the ground surface. The cut is a lateral straight line rather than an arc that may be tipped vertically. The cut is also not sensitive to the grass height, density, or thickness. On the minus side, reel mowers are sensitive to sticks and culls that may litter the lawn. The prospective user of a reel mower clears the lawn of these obstacles before attempting to cut it. Finally, from field tests we know that the energy consumed by a reel mower is the least of the three types. A rotary mower expends most of its energy blowing air, making noise, and not cutting grass, although it is known that the vacuum effect of the spinning blade tends to straighten the grass to its maximum height in some circumstances. The string trimmer is next in energy needs, but again, high tip speeds are needed to be effective. This fact will be important for designing a battery life sufficient to mow a reasonably large area of several thousand square meters. Details of the design of such an automatic reel mower will be given in Section 5.8.

Included in the term "platform" is the software language that will embody the many functions yet to be examined, as outlined in Table 4.2. Of the many choices possible, we selected C for several reasons. First, it is well-known by most programmers and therefore our "product" should be easy to understand by most readers. Next, it is a strongly typed language making authorship of code segments less susceptible to obscure bugs. In this vein, we have deliberately avoided the most intricate and unforgiving aspect of C, namely pointers. The speed of the modern processor is so high that compiled C for our control purposes need not be optimized for speed or memory space through the use of pointers. If one were to

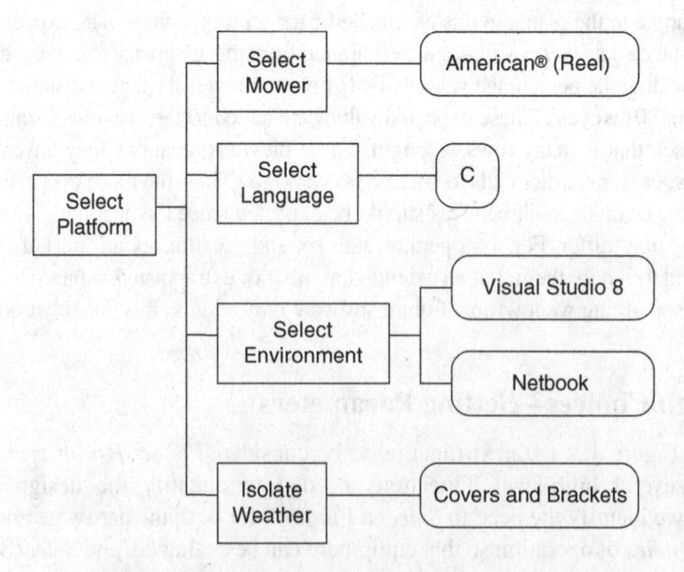


**Figure 5.26** The *Select Platform* function diagram

adopt embedded C-code, on the other hand, memory space and speed may become critically important and the use of pointers more appropriate than not. Next, the C language takes us closer to the hardware than higher-level languages which may become important for embedded applications. Finally, C is available widely and used by vendors in supporting their products, whether using Windows, Mac OSX, or Linux.

In order to *Select* a *Platform* in more detail, we identify the subfunction *Select Environment* which will narrow our options and suggest strategies for development. We have selected Visual Studio 8 Express running on a "netbook" type of laptop. The former is free for the cost of registration, includes a very wide assortment of software development tools, enjoys a wide user base with on-line forums, and is relatively stable in the quickly changing field of computer systems that seem to constantly need "upgrades." The netbook laptop was a generic choice since its small footprint could be either mounted to the robot or set in place at an operator station equally well. It was also the lowest cost option for the greatest processor value, complete with USB ports.

We would emphasize here that the choices we made are in no way binding on the outcome of any field robotic endeavor, including this one. Only a small number of software decisions were based on the platform chosen, and any of the C-code could, of course, be readily expressed in other languages. The lessons learned that are expressed in detail here are applicable to a wide range of field robotic endeavors, independent of the detail choices made.

Finally, the mechanical platform of the lawn mowing robot may need to be protected from the weather. Delicate electronics reserved for laboratory use are indeed found "under the hood" of our robot. It is important to recognize for this purpose that the weather may manifest in *upward* moving wet grass as well as generally downward rainfall. Routine sealing of sensitive circuit elements can become a major cost driver for truly all-weather or underwater applications. We meet our cost target by electing to not take this step in the design process, but rather construct a simple shield for the circuit boards, as shown in Figure 5.27.

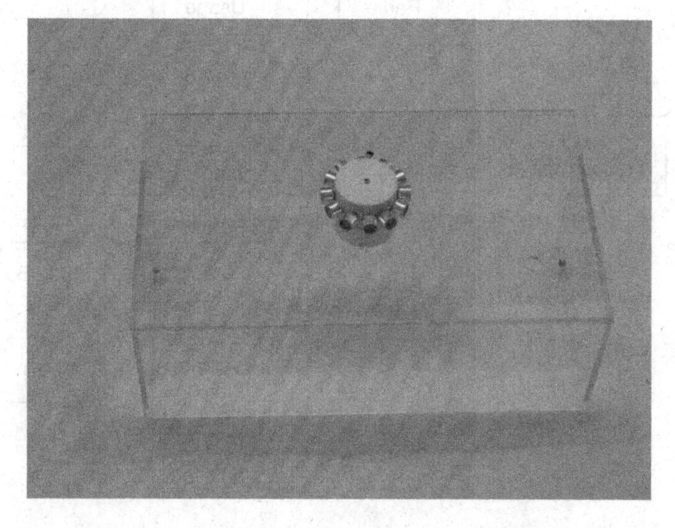

**Figure 5.27** The *Isolate Weather* cover for the field robot

## 5.7 *Select Frequencies*

From Figure 5.28 we are reminded that two very different kinds of frequencies need to be selected (radio and audio), but at the functional level there is great commonality. Both need to be based on the range of operation and the radiated power. For the radio link between robot and fixed station, a minimum of 500 m should be selected so that even distant lawns can be mowed, and a radiated power level of about 1 W is sufficient for this purpose. These values need to also take into account the local obstacles to radio transmission, such as buildings and trees. As a general rule, the higher the frequency, the more important "line-of-sight" transmission is necessary. This is why many cell phones cannot operate in deep tunnels or under large buildings. Another critical component of radio selection is the fact that many modern digital systems send audio information in digital "packets" that may have no relationship to the actual timing of the signals sent. Such a system could not perform the sensitive time-of-flight measurements we have chosen. We normally do not notice the significant time delays found in cell phone technology, since precise timing is of no importance. The opposite is true of GPS, where precise timing is its greatest strength, but its resolution is far too coarse for gardening.

Another consideration for radio functionality is its location within the rules of the Federal Communications Commission. A small number of frequency "bands" are dedicated to public use, and these tend to be overloaded with radio traffic. A public band dedicated to remote ground vehicles is needed[1], and that band is 75 MHz. Please see Figure 5.29 for details. A

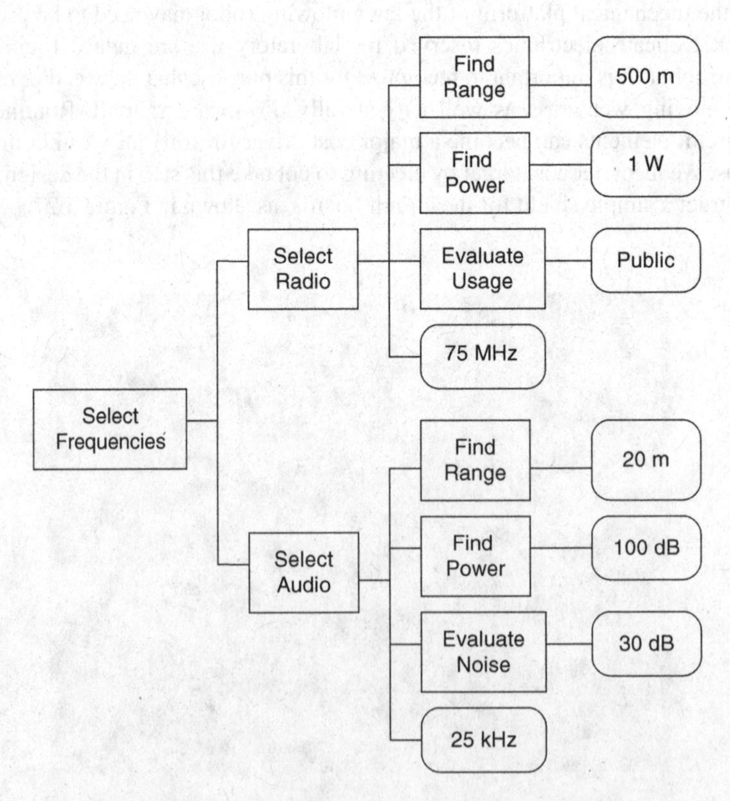


**Figure 5.28** The *Select Frequencies* function diagram

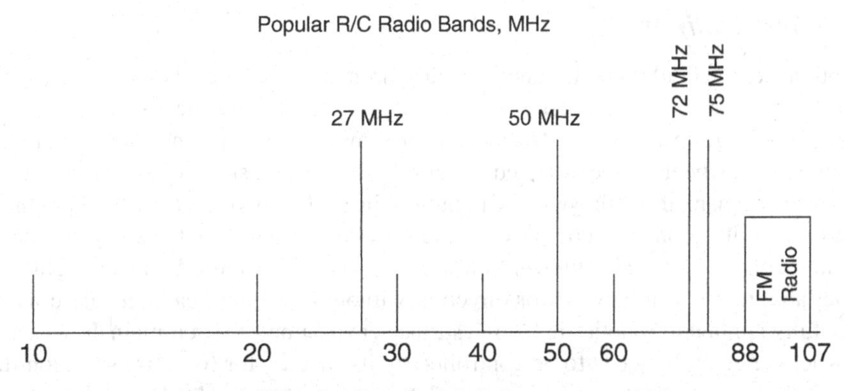


**Figure 5.29** Public radio bands for remote control

salutary effect is that this band is also considered "short-wave" and easily penetrates buildings. Within the 75 MHz band are dozens of separate channels for operating many vehicles in the same area at the same time without overlapping and confounding the control signals. We recommend a thorough study of the rules and methods of R/C technology to make these points clear. This information is separate from the standards employed to send information over the selected carrier wave, as shown in Figure 5.5.

In the audio spectrum, there are practical limits for field robotic applications. Two of these are as follows: at the low frequency end (bass or sub-bass sounds), the rate of information transfer is limited. This band is widely used in nature by elephants and other animals to communicate over long distances, since low frequencies carry well through the air (according to ISO9613[2]). At the upper end, high frequency transmission of sound in air is greatly attenuated. The most common reflective "echo" location devices employ a 40 kHz sound carrier, which can be effective up to about 10 m before the ambient noise reaches the level of the signals sent. While there exists technology to filter signals that are much lower in amplitude than the noise around them, their use is limited to "reconstructive" techniques that do not satisfy the microsecond timing needed for audio time-of-flight measurements.

Our approach to this problem was to sample the ambient noise at a wide range of frequencies from infrasound through ultrasound. We found that a bass drum microphone was ideal for this purpose. We also found that an ultra-low noise balanced input preamp with a gain of about 60 dB was also indispensable. Recall that the decibel measure of a signal is given by:

$$\mathrm{dB} = 20\log_{10}(\mathrm{signal}/\mathrm{noise}).$$

Thus 60 dB is the same as a voltage gain of 1000. The results can be summarized by simply noting that while appliances like hedge trimmers and lawnmowers do generate noise in the 25 kHz range, in general the ambient noise level is only about 30 dB as measured, and much higher in other ranges. Fortunately, the attenuation in air of a 25 kHz acoustic wave is about half of a similar 40 kHz wave, and our goal of a 20 m range can be met with carefully designed receivers. (Please refer to Figure 5.13.) The radiated power needed to reach 20 m reliably was found to be about 100 dB and is readily met by commercial/industrial transducers.

## 5.8 *Select Motions*

The options for the field robot designer are fairly limited, since bipedal (and even multi-footed "beam" robots) are today far too expensive and/or slow for our needs. Figure 5.30 gives the function diagram for *Select Motions.* We examined four options: Ackerman steering, omnidirectional wheels, three-wheeled caster robots, and skid steering. Ackerman steering is the kind of mechanism employed in an automobile or four-wheeled lift truck. While quite maneuverable, its geometric complexity requires path planning routines beyond our scope. Omnidirectional wheels (also known as Mecanum wheels[3,4], Figure 5.31) are highly maneuverable, able to move in any vector direction without changing heading. These have been successfully applied to robotics in benign surface environments. Their main drawback is that each wheel effectively needs to be controlled by its own motor (or a transmission that can create similar speed effects among the wheels). The use of such wheels for outdoor application with grass clippings and other lawn litter in abundance, raises reliability issues. Each passive roller needs to be free of obstructions at all times.

The three-wheeled caster-type geometry is often the basis for small competitive robots, and commercial "random pattern" mowers and carpet cleaners. It also is manifest in the much

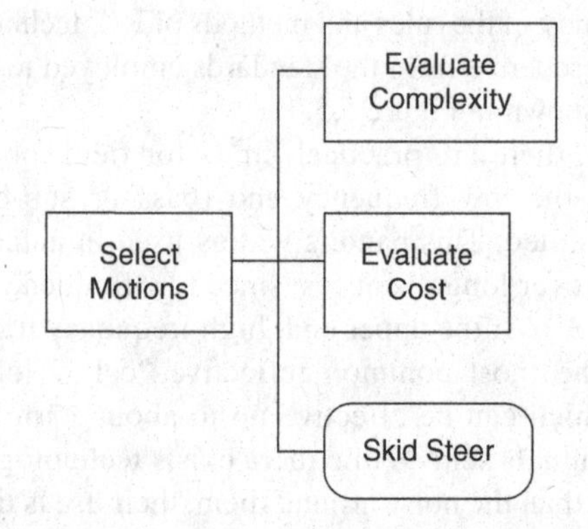


**Figure 5.30** *Select Motions* function diagram

**Figure 5.31** Omnidirectional wheels on a CMU-URANUS robot. (Photograph courtesy of Robert Podnar)

larger lawn tractors in wide use, when one considers the sliding motions of the front wheels as functional casters. An advantage of this design is its simplicity and stability, always forming a kinematic triangle under its center of mass. Its primary advantage is also its own disadvantage: it is very sensitive to turning when confronted with uneven terrain. In a manually driven system, this sensitivity is negated by the exquisite control that people can employ when using such vehicles, as well as the wide swath available.

The four-wheeled skid-steer design has been in use since the beginning of automated vehicles decades ago. It is stable and can "average-out" the dimples and bumps in the typical lawn. It is simple to design and build since only two motors are needed, and both wheels on each side can be simply made to turn together. Its primary disadvantage is that in turning, it can disturb the turf unacceptably. For this reason, the control for such a robot needs to limit the effective turning radius to a relatively large multiple of its width. As Figure 5.32a–d shows, we have selected a skid-steer design despite this disadvantage in favor of its low cost and

**Figure 5.32** Field robot for *Mow Grass*: (a) front; (b) switch side; (c) board side; and (d) back

simplicity. We explore this choice more deeply with other functional considerations in Section 6.5. Other four-wheel-drive designs, such as Sandia Laboratories' "Rattler" were not considered due to the difficulty of incorporating a mower with that type of kinematic suspension. Any dynamic suspension, as found in an automobile, would not guarantee the cut height required.

## Problems

5.1 Show that the gain of the LM358 in Figure 5.6 is equal to 2. Hint: check the LM358 data sheet; see Appendix D.
5.2 Explain how and why the CD4013 of Figure 5.6 "divides by two."
5.3 Suppose there were no programmable digital chips available. How could Figure 5.6 be realized?
5.4 Suppose the +5 V supply of Figure 5.9 was to vary by 10%. What affect would this have on the sound sender? Hint: consider the CD40106 data sheet; see Appendix D.
5.5 Consider Figure 5.19 and sketch the signal for channel B.
5.6 Consider Figure 5.21 and determine the distances separating the robot sender and the two receivers.
5.7 Consider Figure 5.23 and explain why the "`atan2(Δy,Δx)`" function is preferred over the basic "`atan(Δy/Δx).`"
5.8 What is the size of the lawn patch to be mowed according to the plan of Figure 5.24?
5.9 Suppose the 75 MHz band was no longer available by Federal Communications Commission (US) rules. What could take its place? What changes in design would be needed at the functional level?
5.10 What timing accuracy is needed for an audio time-of-flight to be practical? Hint: what is the noise level that would begin to dominate its operation?
5.11 How would the decisions of this chapter change if the budget was doubled? At what cost point could radical changes be made?
5.12 What is the principal reason that a three-wheeled caster-type geometry is much more sensitive to differences in wheel velocity (left vs. right) than a skid-steered design?

## Notes

1. http://www.vantec.com/FCCregs1.htm, especially page 1.
2. ISO9613-2: 1996 (1996) www.iso.org/iso/home/store/catalogue_tc/catalogue_detail.htm.
3. Ilon, B.E. (1975) US Patent 3,876,255, 8 April 1975.
4. http://en.wikipedia.org/wiki/Mecanum_wheel (accessed June 9, 2014), especially CMU's "Tessellator" robot with Mecanum wheels, designed in 1992 for servicing Space Shuttle tiles.

# 6

# Operate Robot

Three subfunctions are shown in Figure 4.3 that distinguish this aspect of field robot functionality from others. Two of them are relatively trivial, but the other will consume the balance of this volume. Figure 6.1 shows the first two subfunctions taken together: the need to *Transmit* and *Receive Signals* via a radio link, and across the system through the USB. As mentioned in Chapter 5, the 75 MHz radio control (R/C) band was chosen for several reasons including its ability to penetrate houses, its ubiquitous availability, and its large user base. The USB was selected for intra-system communication because it persists in the computer field after the demise of the serial port and parallel port of just a decade ago[1]. We may find that the USB will be supplanted as well in the near future, just as it eclipsed "firewire,"[2] and need to be wary of designs that make use of its internal operations. To advance the art of field robotics, we must make a choice, and wireless communication devices are not yet as varied and capable as we need. The market-driven availability of computer equipment dictates that the virtual world supersedes the real, and that communication is more important than control of all but a few devices (printers, scanners, cameras, etc.).

Finally, there must be an instantiation of such control through the USB, and that can be realized by several vendors of "active control." We have chosen the vendor with the most varied and capable assortment of products: Phidgets Inc., Calgary, Canada. For us to use their equipment, one needs only a modest C-library of functions (or alternatively, almost any other computer language in wide use). This library is simply termed `phidget21.h`, is freely available on the internet, and is supported by a substantial user group. We will be using only two of their many products: a servo board (#1061) designed to the R/C standard, and an encoder board (#1047) to interpret time-of-flight signals. The one additional subfunction, *Control System* in Figure 6.2 rounds-out the Figure 4.3 functional map of our field robotic objective: *Mow Lawn.*

*Practical Field Robotics: A Systems Approach*, First Edition. Robert H. Sturges, Jr.

Companion Website: www.wiley.com/go/sturges

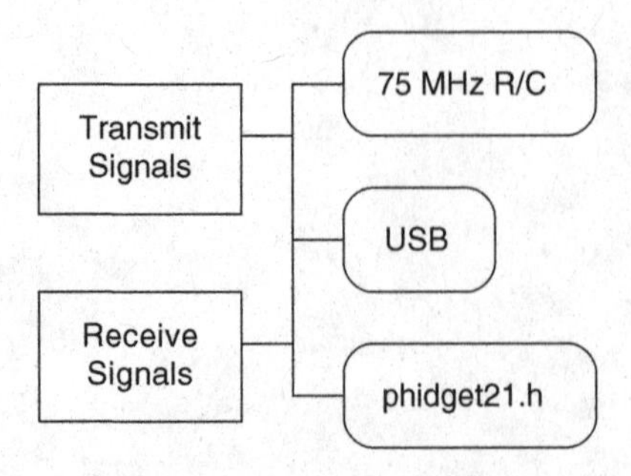


**Figure 6.1** Function diagram for *Transmit* and *Receive Signals*

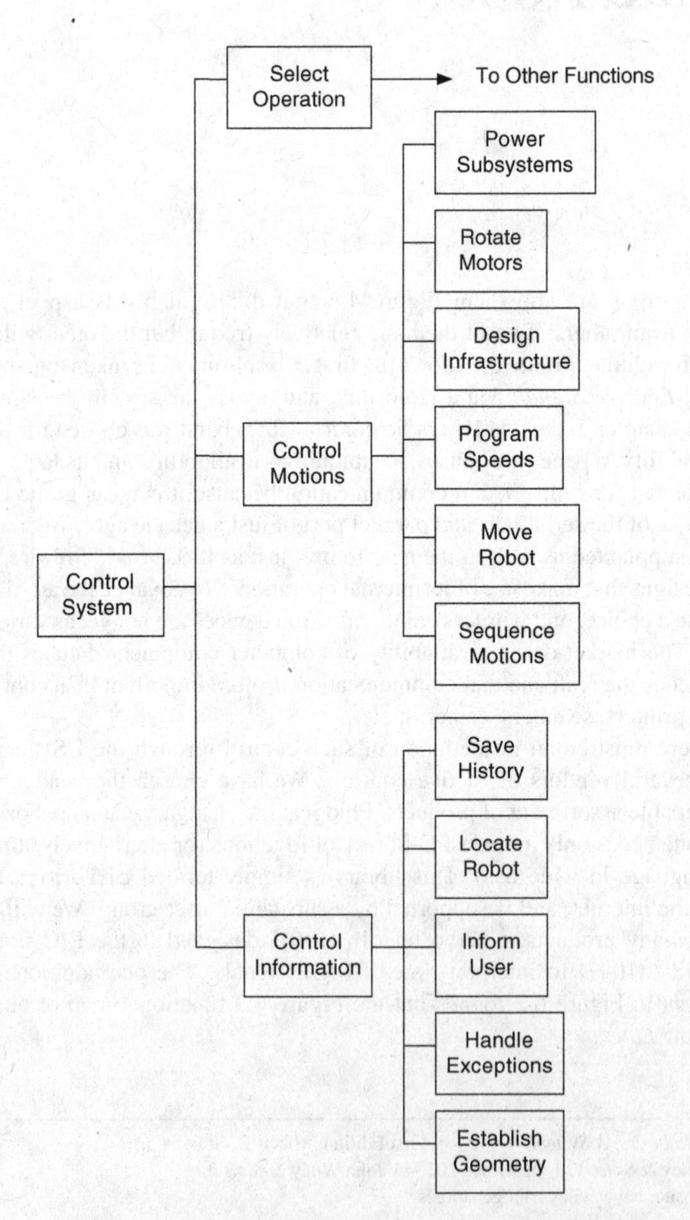


**Figure 6.2** Function diagram overview of *Control System*

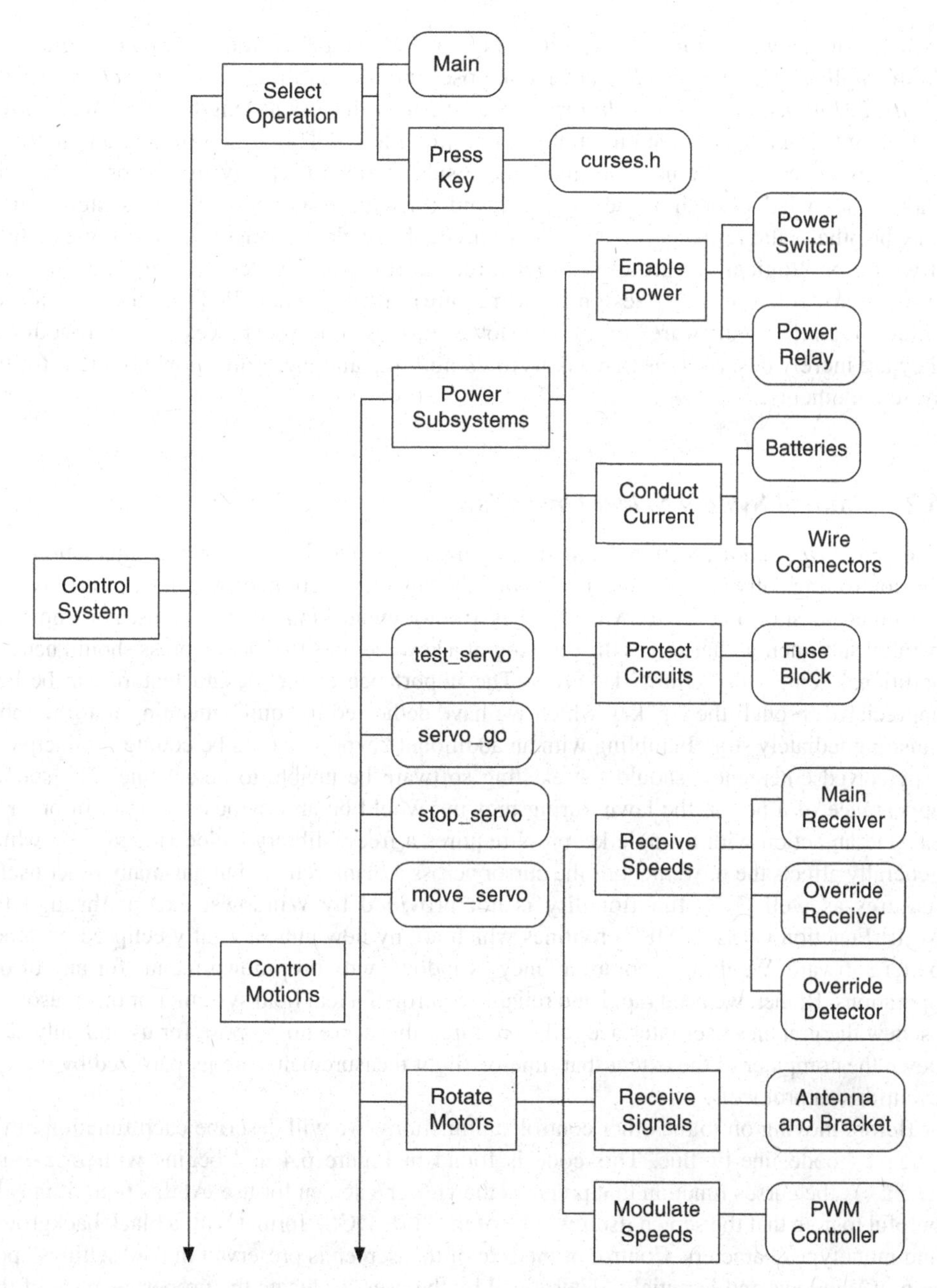


**Figure 6.3** *Select Operation* and other details of *Control System*.

## 6.1 *Control System*

As noted in Section 1.2, the word "control" is used here as an active verb. It instructs us to further decomposition into the many subfunctions of the system that integrate hardware, software, and operator input. We avoid the verb "integrate" since it really does not tell us what to do specifically, except to somehow design and build a system. We will be far more explicit. In fact we begin our discussion of control with the function diagram of Figure 6.2. At the highest level, *Control System* decomposes into three subfunctions: *Select Operation, Control Motions,* and *Control Information.* One may discuss at length the choices shown in Figure 6.2 and debate the hierarchy of function blocks. This is helpful and a potentially innovative exercise. We have arranged these subfunctions in a way that helps explain the "how" and "why" of each of the hardware and software components in our system. There may be other valid representations, and we invite the reader to consider a few. For example, it would be "logical" to group the first three subfunctions of "control motions" (power subsystems, rotate motors, design infrastructure) into a group called "hardware" and the latter three into a "software" grouping. However, these one-word titles are *not* functional, they are merely descriptions that ask us to *do* nothing, and therefore show no intent for the system elements.

## 6.2 *Control System Select Operation*

The *Select Operation* function detailed in Figure 6.3 explicitly calls for user interaction with the controlling software. In typical "C-style" the "main" function, of which there is only one, is limited to start-up, shut-down, and pick-from-a-menu kinds of functions. An important parallel subfunction that serves the pressing of a key, requires that no keypress should need an additional "enter" or "return" keypress. The importance of this design feature can be best appreciated through the "q" key which we have dedicated to "quit," meaning that the robot must immediately stop. Fumbling with an additional keypress would be counterproductive to a perceived emergency, should the existing software be unable to detect one. The sudden appearance of a pet on the lawn during mowing would be an obvious example. In order to execute an action with a single keypress requires a free C-library called `curses.h` which generally affects the movement of the cursor across a blank screen, but has many other useful features as well. This functionality is not provided by Windows, except through the Multi-Function-Classes (MFC) routines which are by now almost totally eclipsed by other, better software. We do not require a fancy "window" with drop-down menus for any of our operations. Rather, we need rapid and reliable control of a real field system. For this reason we eschew the graphics user interface (GUI) routines that serve no purpose for us and only slow down the computer to the extent that time-of-flight measurements are jeopardized by mouse-event-driven protocols.

Before moving on to the other control subfunctions, we will describe each function of the `main()` code line-by-line. This code is found in Figure 6.4 and begins with `screen_init()`, the curses function that prepares the viewer's screen for use. At this point it may be helpful to note that the screen itself will be of the "MS-DOS" format with a black background and monotype characters. Control of the size of this screen is preserved in the "settings" portion of Windows and is initially determined by the user. To locate the properties page of this window, the user must click on the little "`c:\`" icon in the upper left corner of this window. We have extended the window to nearly the full size of the computer screen, but left a small margin to select other software functions on the Windows desktop.

A word or two about function block representation of software is in order at this point. We surely could have functionally decomposed all of the routines we have already described (and will be detailing) rather than following the usual "flow chart" description

```
int main()  {

     //int i, j;                                    //for bug test
     //for(i=0;i<12;i++) for(j=0;j<4;j++)  dist_stack[i][j] = 99;
      screen_init();                              //start the screen
      open_encoders();                            //make sure the encoders are on
      display_Eproperties(ENC);                   //show all the properties first
      open_servos();                              //make sure the servos are on
      display_Sproperties(SRV);                   //show all the properties first
      stop_servo();                               //initialize servo positions to neutral
      update_display3();                          //for testing
      receivers_out();                            //select receiver locations outside
      //receivers_in();                           //select receiver locations inside
      while (doloop) {                            //always true until you quit
            current_getch = getch();              //get a character
            switch (current_getch){              //determine what to do with it
            case 97:                             //'a'  for read average
                     get_mode();                  //read encoders, 4 val's returned
                      break;                          //terminate case
                      case 99:                   //'c'  for curve
                      break;                        //terminate case
                      case 100:                  //'d'  for distances
                      // d_test();
                      break;                        //terminate case
                      case 101:                  //'e'  for encoders
                      break;                          //terminate case
                      case 102:                  //'f'  for feedback
                       break;                          //terminate case
                       case 103:                 //'g'  for go!
                           servo_go();                //move out
                        break;                       //terminate case
                        case 105:                //'i'  for NOTHING
                        break;                       //terminate case
                        case 108:                //'l'  for locate
                           locate(0,0);          //read encoders, 4
                        break;                   //terminate case
                        case 109:                //'m'  for move
                        move_servo();//move servos, 2
                        break;                   //terminate case
                        case 112:            //'p'  for point_list
                              fileread();    //get a gauntlet on file
                        break;                   //terminate case
                        case 113:            //'q' for quit
                              stop_servo();  //whoa!
                              doloop = 0;    //terminate do loop
                        break;                   //terminate case
                        case 114:                //'r'  for read file
                              fileread();
                        break;                   //terminate case
                        case 115:                //'s'  for stop
                              stop_servo();//init servos to stop
                        break;                   //terminate case
                        case 116:                //'t'  for time
                              GetSystemTime(&time);
                              millis = time.wMilliseconds;
                                //secs = time.wSecond;
                                printw(" %u ", millis);     //what time is it??
```

**Figure 6.4** The C-Code for main()

```
        break;                          //terminate case
        case 118:                   //'v'  for velocity test
        //  v_test();
        break;                          //terminate case
        case 119:                   //'w'  for waypoint
             waypoint_drive();//get xd, yd, & speed from list
        break;                          //terminate case
        case 120:                       //'x'  for whatever experimental
             test_servo();
        break;                          //terminate case
    }                                   //this is set up for more cases
      }
    printw("Closing...\n");                    //we quit, so tidy up the house
    CPhidget_close((CPhidgetHandle)ENC);             //close the phidget
    CPhidget_delete((CPhidgetHandle)ENC);            //& delete the object
    CPhidget_close((CPhidgetHandle)SRV);             //close the phidget
    CPhidget_delete((CPhidgetHandle)SRV);//& delete the object
    screen_end();                              //all done, exit
    printf("TEST ENDS\n");                     //tell the user
    return 0;
    }
```

**Figure 6.4** (Continued)

and explanations. We have employed the flow chart approach for three reasons: first, we decided that our readers would expect this format; secondly, in "C" the functionality is to some extend already expressed by the {...} structures; and thirdly, we would have needed to reverse-engineer these code segments to develop their functional representations since they actually were not coded with function diagrams determined initially. We would note that coding of software systems of a higher level of sophistication than needed for this robot would be aided greatly by designing the code by function representation, that is, Value Engineering[3]. This step should involve the many parties (managers, designers, coders, and users) in a meaningful dialog about the code sub-functions *before* coding begins. In this way coding efficiency can be enhanced and quality more likely assured before testing takes place.

## 6.3 All About `main()`

Invoking `open_encoders()` follows for the purposes of setting up the USB connection to the encoder board in the "USB translator" and displaying to the user that this connection was successful. Alternatively, and error may be reported. In this case, the user would be advised to check the connections and try again. With the encoder connection made, `main()` then invokes `display_Eproperties (ENC)` to show the user the standard information regarding serial number and type of encoder board now attached. The formal parameter `ENC` is declared in the globals and declarations list near the head of the program. It stands for the single encoder board in use. Early in the conception of the *Mow Lawn* system we considered communicating with up to 12 separate encoders on 3 boards (4 on each). As the system design matured, we realized that one would only need two receivers at a time for robot localization. If the original intent (of a dozen receivers) is desired, the `ENC` declaration would be duplicated twice and relabeled for `ENC1`, `ENC2`, and so on.

The next step in `main()` is to open the servo board by invoking the `open_servos()` function. Similar to the previous open statement, the successful opening of the Phidget Advanced Servo Board results in a short display of serial number, version, and channel numbers for this board. If the board were not functioning or unplugged, an error would be reported. The condition of the servo commands are not guaranteed upon opening, so the very next statement to `stop_servo()` appears. This function sets each servo channel to its "center" or "off" position.

With the preliminary set-up completed, `main()` invokes `update_display3()` to show the user a short list of the x, y, and z values collected. This list shows the history of the last 10 readings gathered by the `locate()` function. These will be zero on startup if no errors have occurred. A set of constants needs to be established for each of the receiver locations. Since this system has been operated both indoors, during development, and outdoors in practice, there are two such sets locating two different sets of receiver locations. A simple function is invoked to establish these values, `receivers_out()`. A similar function is commented out, but available for `receivers_in()`.

The principal part of `main()` is then initiated with a `while` loop. This loop is dependent on the global variable `doloop`, which is initially set to a "1," or `true`. Should any function command a "quit" of the system, this value would be set to zero, and the `while` loop terminated. Barring that event, the first task to be done every time is to collect the most recent keypress. As mentioned earlier, the `curses.h` library ensures that no "enter" key is needed to recognize the specific request for a system action. A C-type `switch` on the character value is then specified, passing control to a number of functions designated by a corresponding single letter. Many of the software subfunctions used by the system code are available for individual invocation, as can be seen by a quick scan of the cases in the while loop. With no more commands, or a "q" for quit, `main()` then ends with a formal shut-down of the encoder and servo boards, and a short message to the user.

The *Select Operation* has been described, so it is time to discuss the next of the three subfunctions of *Control System: Control Motions.* Six subfunctions plus several simple software functions are needed to carry out *Control Motions,* as seen in Figure 6.2. The medium of a single sheet of paper cannot hold all of them, so these functions have been arranged in several consecutive figures. These will now be addressed from top to bottom of the function diagram.

## 6.4 *Control System—Control Motions*

The first of these is *Power Subsystems,* and we need to do three things to realize this fundamental concept: *Enable Power, Conduct Current,* and *Protect Circuits.* Again, these may seem obvious in hindsight, but we stressed that every component in the system needs a *raison d'être,* and there they are in Figure 6.3. The detailed electrical schematic for these functions is shown in Figure 6.5. Few of the circuits are completed in this drawing, but will be in functions and schematics to be described shortly. The choices for these components are based primarily on the allocations made earlier in *Set Parameters,* but there is a wide latitude for specific parts in several instances. For example, the net voltage and amp-hour ratings for the batteries are set by the choice of motors and the desired operating time for the system, that is, the size of the lawn to be mowed. We have selected a pair of 12 V, 28 AH sealed lead acid (SLA) batteries so that nearly 3 h of operation are possible under heavy load. The motors needed to drive the reel-type mower will be discussed further, but their net power rating should be mentioned here.

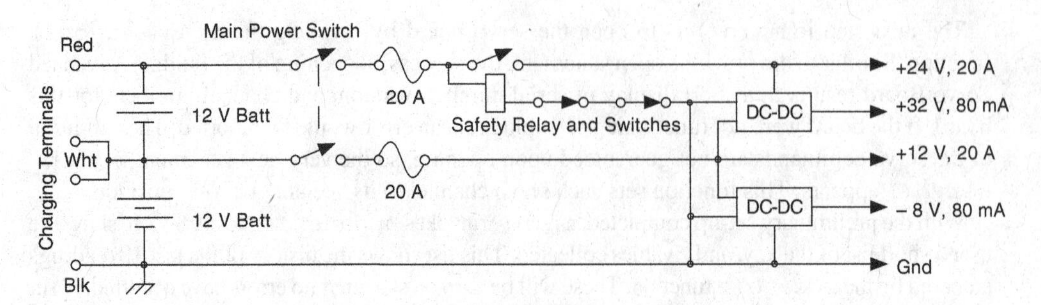


**Figure 6.5** Schematic details of *Power Subsystems*

Each motor should be a minimum of 100W shaft output in order to develop the speeds needed for successful mowing. Under heavy load, one should expect a current demand of 8.3 A from the drives. Taken together, the batteries and motors comprise about 90% of the system total weight of 356N. Although not specifically mentioned as an allocation, this weight must be small enough to permit autonomous operation on slight grades. We have tested the system on a 5% grade with no difficulties in current draw or slippage. We do not recommend greater grades without careful study of the tractive ability of the tires selected. This study is left as an exercise.

In order of appearance in Figure 6.3, the Power Switch is a 20 A capacity DPDT, so that both the 12V and the 24V leads of the batteries can be enabled or disabled together. The 12V lead serves as the "ground" lead for the *Amplifier Board* under *Send Sound,* shown previously, but the most negative lead of the system, the negative lead of the 12V battery, need not be switched. A DPDT Power Relay serves to enable the motors and other functions of the robot under the condition that a set of safety switches are in their normally closed positions. These will be discussed further in *Stop Robot* in Section 6.8. A set of "Faston" wire connectors are used to complete the conduction of current to the drive motors. These are high-current connectors, unlike the screw terminals used for all of the custom circuit boards on the robot. Finally, a Fuse Block is inserted after the main Power Switch in order to Protect the Circuits from accidental overload due to shorts or other mishaps. The fuse ratings are chosen at 20 A each to enable the high start current of the drive motors and to enable the system to operate during a stall condition. Stall conditions could occur if the mower reel became jammed on a stick. This condition could be sensed by the sharp drop in forward progress of the robot since we maintain a "log" of the recent x and y positions. An unprogrammed deviation from the expected values (provided that there ARE values) may indicate a stall. Such a condition can readily be corrected by a short "backup maneuver" of the robot to dislodge the stick. Code for this purpose existed in earlier versions than that shown here, but was removed since a stall has never occurred in practice.

Next in line on the *Control Motions* "and" line are four simple functions that may be invoked directly from `main()` or called by other functions. They are: `test_servo()`, `move_servos()`, `servo_go()` and `stop_servo()`. `Test_servo()` requests a number for a servo channel and another number, typically 0 through 200, to be applied to that channel using the Phidget advanced servo calling function. We use it to validate each servo independently during system checkout. The function `move_servos()` similarly requests 2 numbers from 0 through 200 to be simultaneously applied to the motor drive channels. With this short code function, one can test the radius of turn or the straight-line ground speed of the robot during system checkout. This function has proven useful in testing the battery voltage and slope effects on robot speed as well. The function `servo_go()` applies the numeric values appearing in the two global variables `mov[0]` and `mov[1]` to the Phidget function that drives the motors only. The designer of code for a field robot should be aware of the details of the Phidget calling function itself. Three formal parameters are needed to drive any of the eight servo channels on the Phidget advanced servo board. They are: a "handle" or address for the board which is declared early in the code as `SRV`, and to which the compiler will assign a value; the channel number from 0 to 7; and the value sent to the servo board that determines the pulse position of that channel. The foregoing C-functions are sufficiently simple that they will not be discussed in any further detail.

## 6.5 *Control Motions—Rotate Motors*

The function *Rotate Motors* applies to the left and right side drive motors alone. There are at least three subfunctions needed to accomplish this action, and all are located onboard the robot. First, speeds must be received, including zero speed for stopping. To do this, there is an R/C *Main Receiver* set to the same channel as the *Main Transmitter* which is driven by the "USB translator" board, as discussed in Section 5.2. Additionally, a separate R/C *Override Receiver* is set to a different channel matching a hand-held *Override Transmitter.* A manual control mode can be invoked without computer interaction by simply switching "on" the *Override Transmitter.* On board the robot is a custom circuit board, the *Override Detector* (Figure 6.6), that detects the presence of a signal coming from the *Override Receiver* (Figure 6.7). In order to detect the presence of a valid set of drive signals on this receiver, rather than random noise, its channel 1 signals are stretched to 10 ms by a one-shot and then sent to an LM567 "tone decoder." (The R/C convention numbers the channels from 1 to 8 rather than 0 through 7.) This decoder raises a signal line on the board whenever a pulse is received consistently every 20 ms. When so detected, the board switches a DPDT relay to direct the channel 1 and channel 2 pulse widths to the PWM motor controller to be described shortly. The *Override Detector* also indicates with a red LED that this operating mode is in effect.

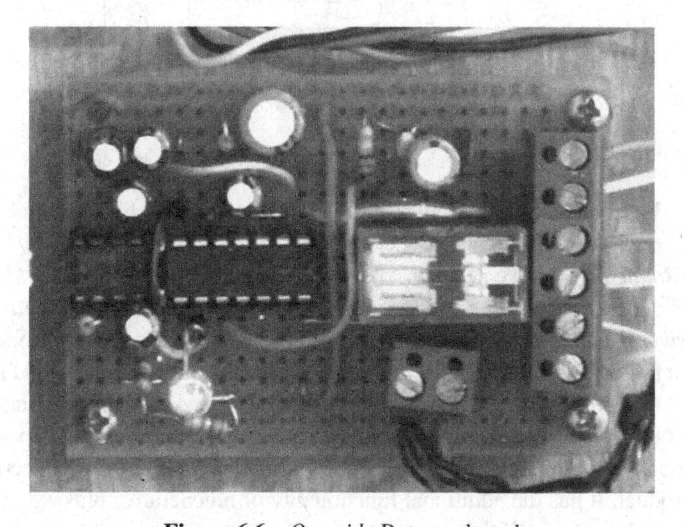

**Figure 6.6** *Override Detector* board

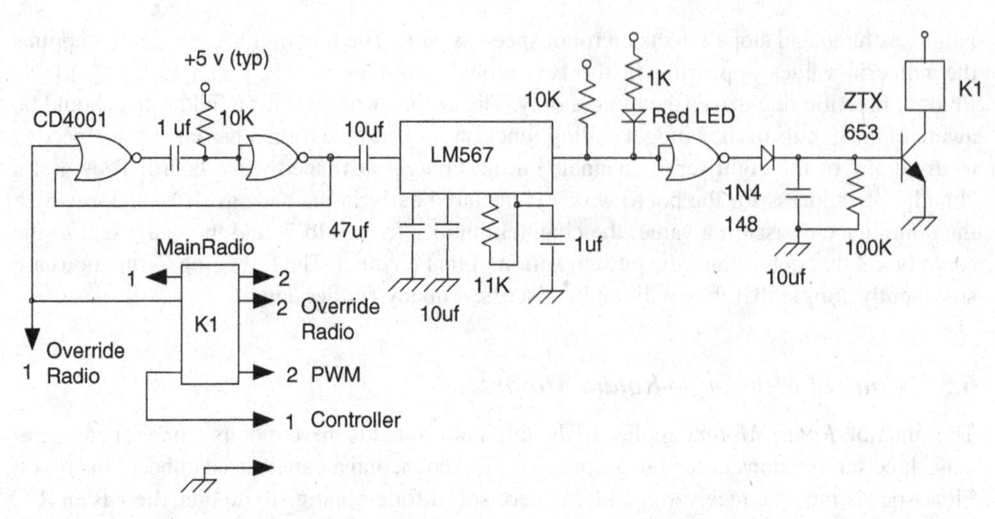


**Figure 6.7** Schematic for the *Override Detector* board

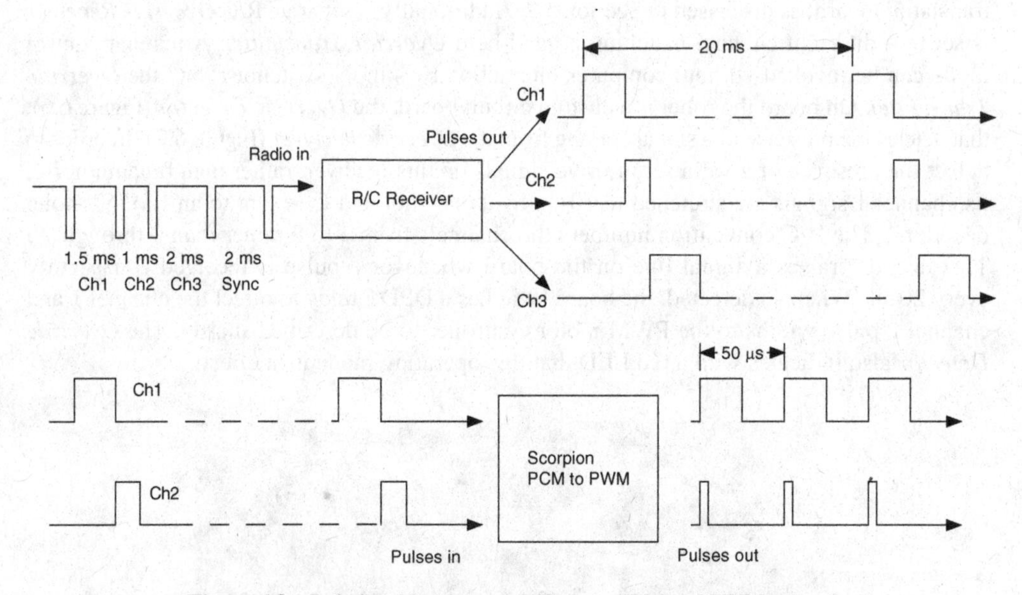


**Figure 6.8** Relationship between R/C servo PPM and PWM control

In order to *Rotate Motors,* we need two other functions: *Receive Signals* and *Modulate Speeds.* At this level of detail, *Receive Signals* is accomplished with the *Antenna and Bracket* artifacts for the *Main Receiver.* A commercial, packaged PWM controller is used to *Modulate Speeds.* By PWM, we mean that a full 24 V is applied to the drive motors, but the duty cycle is varied from 0 to 100% by the received R/C pulse width, as shown in Figure 6.8. A growing number of such R/C-to-PWM boards are becoming available for motor control in competitive robotics, in capacities from a few amperes to over 200 A. We selected a dual 25 A unit from Dimension Engineering called their Scorpion™ product. It has the additional functionality of regenerative braking. A rapid stop can be used to charge the batteries if the motors begin to act as generators in slowing down the robot.

## 6.6 *Control Motions—Design Infrastructure*

Mechanical design skills are needed in Practical Field Robotics, especially to accomplish the *Design Infrastructure* subfunction of Figure 6.9. In this group, we have included seven vital mechanical support details. These comprise the following: *Complete Circuits* which is done

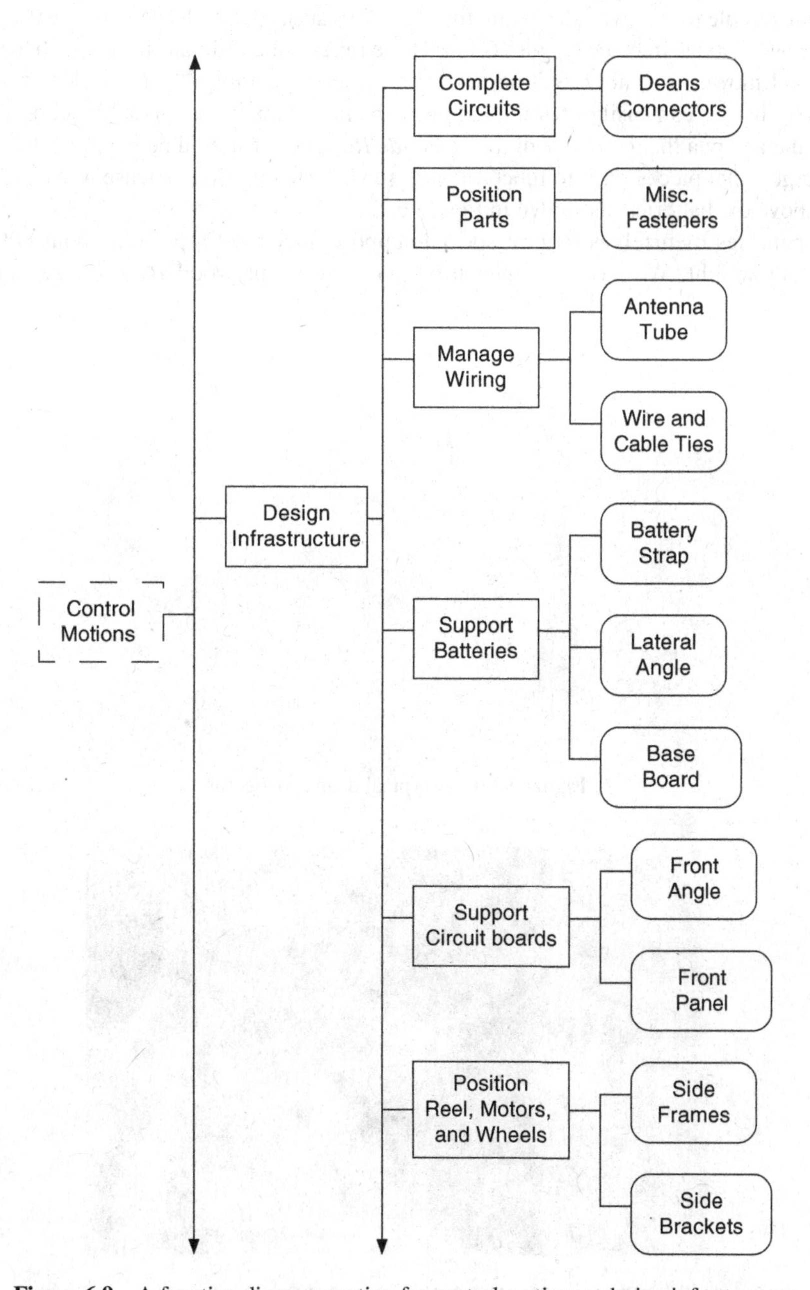


**Figure 6.9** A function diagram portion for control motions—design infrastructure

by a popular type of R/C connector with three standardized leads called a Deans Connector™, as shown in Figure 6.10. These connectors are needed to interface with the two radio receivers on board the robot. *Position Parts* is the functional way of specifying any number of otherwise unremarkable brackets and fasteners. When there are specific position needs that may have critical allocations (strength, stiffness, precision, etc.), more detailed functions can be called out. For example (as shown later in the function diagram), four Nylok™ nuts are used to hold the four wheel axles in place (Figure 6.11). These replace the ordinary fasteners that are a part of the reel mower assembly as it comes from the manufacturer. Next, we need to *Manage Wiring* so that we can easily identify and place the more flexible parts of our robot. One such wire is the antenna that comes with the *Override Receiver.* It should be positioned away from other large metal pieces for it to function well, so we have elected to encase it in a plastic tube taped above the batteries, as shown in Figure 6.12.

The batteries themselves require special support since they represent about 80% of the total robot weight. We have mounted them on a sturdy plywood *Base Board* (measuring

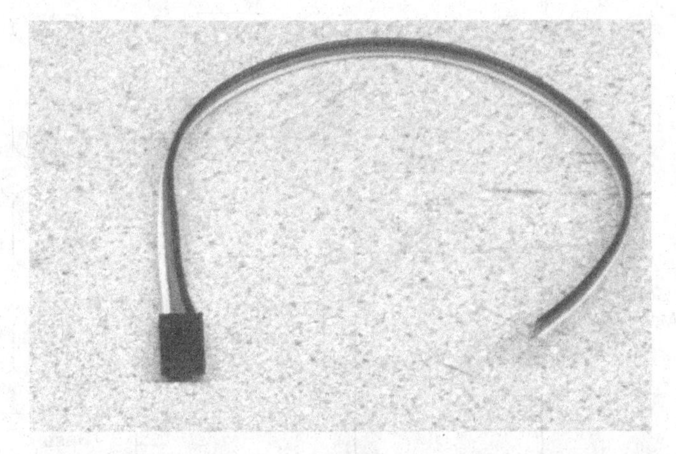

**Figure 6.10** A typical deans connector

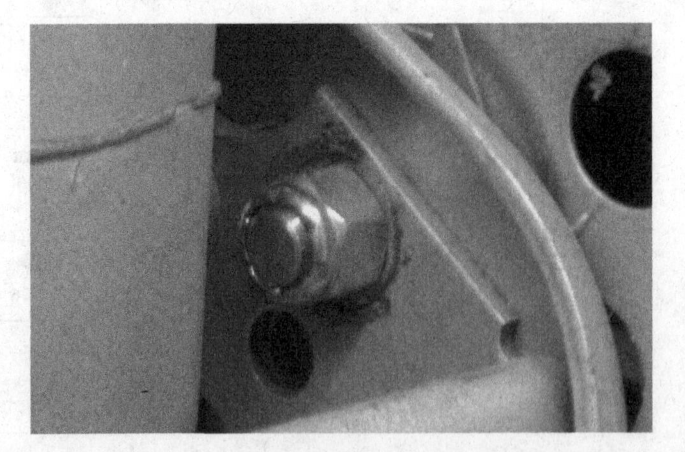

**Figure 6.11** Nylok™ nuts needed for wheel shafts

**Figure 6.12** *Manage Wiring* tube for a receiver antenna wire

**Figure 6.13** Detail of support for *Front Panel*

12 mm × 228 mm × 508 mm). We maintain its location with a lateral "slotted angle" piece and a metal *Battery Strap,* also shown in Figure 6.12. Several custom and commercial circuit boards already mentioned are fastened to a plywood *Front Panel* measuring 5 mm × 127 mm × 355 mm. It, in turn, is held perpendicular to the base board by an aluminum extruded angle piece. It can be seen at the junction of the two boards in Figure 6.13. Also notice the two DC–DC converters on the little board in the center of the photograph. We first saw these in Figure 6.5.

A crucial bit of mechanical design is embodied in the function *Position Reel, Motors, and Wheels,* each of which must be held rigidly in place with respect to each other. We have constructed the bulk of our robot frame using the *Side Frames* from two identical reel mowers. One of the side frame pairs retains its original reel assembly and adjustable cutting bar, plus its wheel holders and mud guards. The original mower from which this assembly was taken is shown in Figure 6.14. Notice that the reel shaft is driven by a pair of pinion gears that are in turn driven by an annular gear molded into each wheel. In this way, the speed of the reel becomes a factor of 4.3 faster than the drive wheel, as seen in Figure 6.15. Not shown is a sliding pawl built into each reel drive gear, so that only the faster of the two drive wheels engages the reel. When in straight-line motion, both pawls engage and both small gears drive

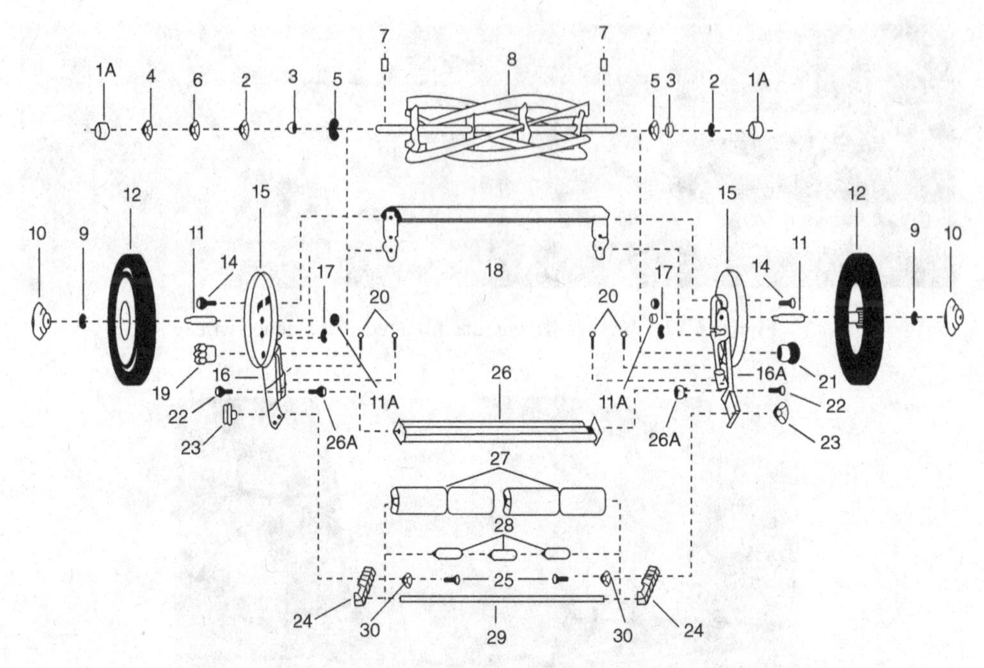


**Figure 6.14** Original American™ reel-type mower diagram

**Figure 6.15** Gear drive for reel mower

the reel shaft. American Mower™ also offers wheel sets and metal drive gears as replacement parts. We have deleted many other unneeded parts from Figure 6.14 and substituted new, longer axles in place of the original ones (Figure 6.14, 11).

A second pair of side frames was inverted and bolted to the first set of side frames with two *Side Brackets* shown in Figure 6.16. The second reel assembly and both push handles were not used. To drive and support the second set of wheels, the reel shaft was replaced with the

**Figure 6.16** *Side Bracket* for the field robot

**Figure 6.17** *Motor* for the field robot

two output shafts of two drive motors, preferably a left- and right-handed matched pair. These motors each carry a worm gear assembly and bearings as shown in Figure 6.17. Many types of such motors are found on the market today; ours were from Everest & Jennings and intended for use as wheel-chair drive motors. Many newer motors of this size (about 100 W) are in use for competitive robotics as well. At this point in the description, the elongated wheel axels are supported only by the side frames and mudguards of the original mower, plus new Nylok™ nuts. These special nuts are needed to keep the vibration and frictional torque of the reversing wheels from loosening the axles. To *Support Wheels* fully, a set of *Spacer Bars* is needed just outboard of the wheels, as in Figure 6.18. With them are placed a set of washers to reduce wear and friction and an *E-ring* on each axle to restrain the parts axially. With this infrastructure in place, we can begin to examine the three remaining subfunctions of *Control Motions.*

**Figure 6.18** *Spacer Bars* for the field robot

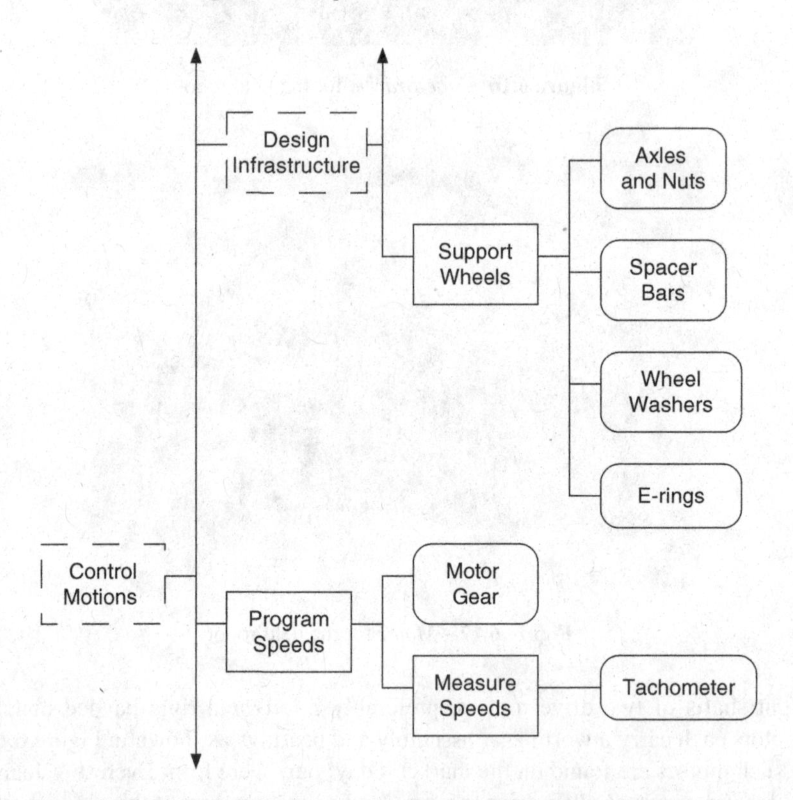


**Figure 6.19** Function diagram portion for *Control Motions—Program Speeds*

## 6.7 *Control Motions—Program Speeds*

A necessary function we used in *Control Motions* was to measure and program the speeds of the drive motors for straight-line and arcuate motions, as shown in Figure 6.19. For this purpose, we equipped each motor shaft with a smaller *Motor Gear,* and used that to drive a *Tachometer* of our own design. We designed our own tachometer since the interior motor

**Figure 6.20** Simple tachometer setup for field robot motors

shaft driving the worm and worm gear was not directly visible, and the output shaft speed was rather low. Instead, we attached a small gear to a permanent magnet DC motor and temporarily clamped it to the drive motor, such that it could be driven by it, as shown in Figure 6.20. A permanent magnet DC motor acts as a generator when connected this way, and we measured the voltage it produced while the robot was running on a test stand. To calibrate the DC motor/generator itself, we drove it with a drill press at a known speed, since the voltage–speed curve is known to be linear.

## 6.8 *Control Motions—Move Robot*

In order to simply move the robot, we need to do at least four things: *Drive Wheels, Skid Steer, Follow Lines,* and *Stop Robot.* The explanations to follow will complete most of the mechanical design descriptions and lead into the software requirements. In some places there will be design latitude, and in others the hardware and software functions will be closely coupled. Figure 6.21 gives the function diagram. First, to drive the wheels, a little more detail is needed from the earlier description. We need to attach the motors so that the shafts align as required, but also in rigid configuration to restrain motor torque. For these purposes, we attach the motors with *Motor Mounts* and *Bolts* (Figure 6.22), align the motors with a *Cross Beam* (Figure 6.23) and drive the wheels with the same type of gears already used to drive the reels of the other pair of wheels in the four-wheel assembly. That is, we drive the annular gears inside each wheel with our drive motors, but instead of pawls, we attach these gears with two set screws each. We position the set screws at 45° from each other on the gear periphery so that the maximum holding torque is achieved. The *Motors* themselves finally appear in the function diagram where they belong, in a chain of intentional reasoning that can be traced all the way back to *Mow Lawn.* We reiterate this fact to show that *all* parts and functions form a part of the *intent* from the most prominent to the least noticeable.

Next, in order to *Skid Steer,* we need to add a few new functions and parts. We need to *Contact the Ground.* This is no trivial matter, since a rigid four-wheel device, as this is, may

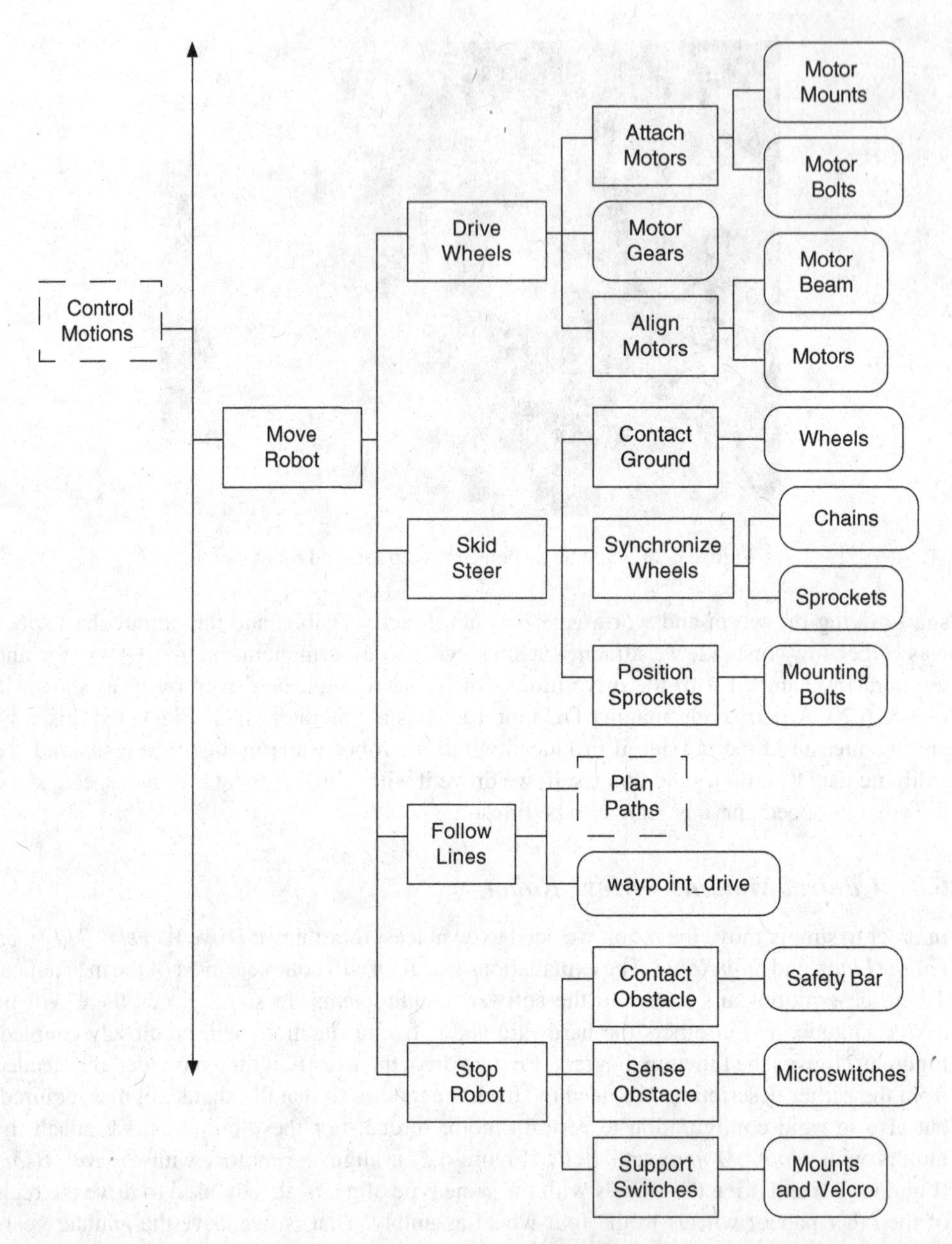


**Figure 6.21** Function diagram portion for *Control Motions—Move Robot*

not have all four *Wheels* on uneven ground at the same time. In fact, we rely on the compliance of the lawn to achieve this mode of guidance for the robot. We have found that the most important disturbance to straight line control and line-following is the unevenness of the turf itself. For this reason, classical control theory cannot completely describe our system. Also the disturbances may be Gaussian (truly random in appearance and magnitude) but they are sufficiently large to render even modern filtering techniques inadequate. Unfortunately, we

**Figure 6.22** *Motor Mounts* and *Bolts* for drive motors

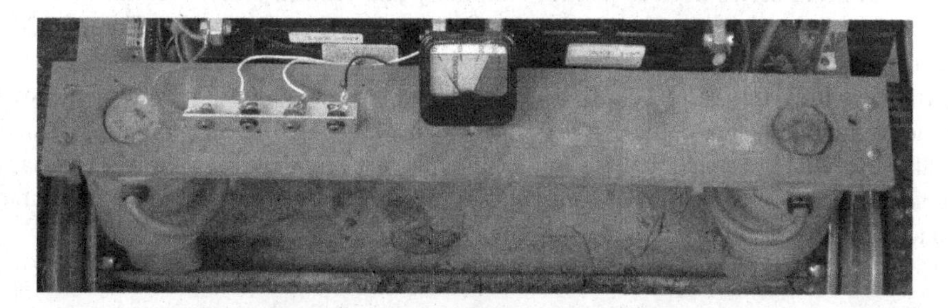

**Figure 6.23** *Cross Beam* for drive motors

have only the wheels and some clever programming to solve this problem, as will be discussed in more detail in Section 6.10.

To achieve the skid needed, we *Synchronize the Wheels* with *Chains* and *Sprockets*. This mechanical arrangement also transfers torque to drive the reel shaft itself, without reliance on ground contact. Naturally, the sprocket wheels need to be concentric with the drive axles. The distance between the axle holes in the *Spacer Bars* mentioned earlier are determined by the diameter of the sprockets and the pitch of the chain. As mentioned also in Section 6.7, there is a need to restrict the radius of curvature to a large value so that the turf is not damaged by the skidding of the wheels. We find that a 10% difference in speed between left and right wheels is readily accommodated by our lawns, but that a 30% difference may leave tell-tale marks, just as pivoting power mower wheels often do.

We have hinted that one way to mow a lawn with a field robot is to set waypoints and *Follow* straight *Lines* between these points. Programming is simplified over arcuate motions, and most curves are readily approximated with a number of short, straight lines. Our intent at this point may begin to differ from alternative designs, such as three-wheel systems with castering. We shall see, however that this very different-appearing approach changes very few of the functional decisions, and even just a few of the mechanical, electrical, and software designs of such a system. The software approach we have taken (`waypoint_drive()`) will be discussed in detail with respect to the next function block, and it will depend strongly on the previous function *Plan Path(s)*.

**Figure 6.24** *Safety Bar* for drive motors

Before delving into our control software from a functional perspective, we need to *Stop the Robot* for safety concerns. The functions we have included comprise: *Contact Obstacle, Sense Obstacle,* and *Support Switches.* We have not elected to use sensors, such as the popular Sharp™ GP2D12 series devices just for simplicity and cost. In order to contact an obstacle, we include a *Safety Bar* reaching across the front of the robot as shown in Figure 6.24. Also shown are the *Microswitches*™ which support the bar, and the switch *Mounts,* attached to the side frames with *Velcro*™. We chose an easily detachable medium for mounting the switches in anticipation of many possible contacts and possible bending of the safety bar. Alternatively, the safety bar could be made flexible and the mounts more rigid. Design choices are many. We have already mentioned that the *Microswitches*™ close a circuit that enables the drive motors. Contact with an obstacle simply disables the drives.

## 6.9 *Control Motions—Sequence Motions*

A portion of the function diagram for *Control Motions* is given in Figure 6.25 to show how to implement the *Sequence Motions* subfunction. As mentioned, the code which does this sequencing is `waypoint_drive()`, and it depends on the geometry of *Plan Path(s).* The details of mowing in this instance was decided to include "patches," or sections of lawn that could simply be described by raster-like motions of the field robot. Each end of a raster line has a waypoint that is located by either measurement, mapping to scale, or by using the robot itself to `locate()` the start and stop locations. The list of waypoints is saved with the speed desired of the robot and the receiver identities needed to navigate through the entire patch. This information is stored in a file and generated by the user using Notewriter™ or similar text file authoring code.

## 6.10 *Control Information*

The software needed to carry out the *Control Information* function is, of course, developed by functional decomposition, as shown in Figure 6.26. From a broad perspective, we identify six subfunctions, which we will describe here. First, we *Save History.* We do this for several variables and for several reasons. The "history" of the robot's motion is preserved in four stacks of data, as shown in Figure 6.27. Each stack is composed of the most recent value followed by the next most recent, and so on. The length of each of the four stacks is common: `stack_length`, and it is chosen to encompass a reasonable amount of time and corresponding number of 80 ms cycles. In the case shown in Figure 6.28, we set this value to 60, so that an

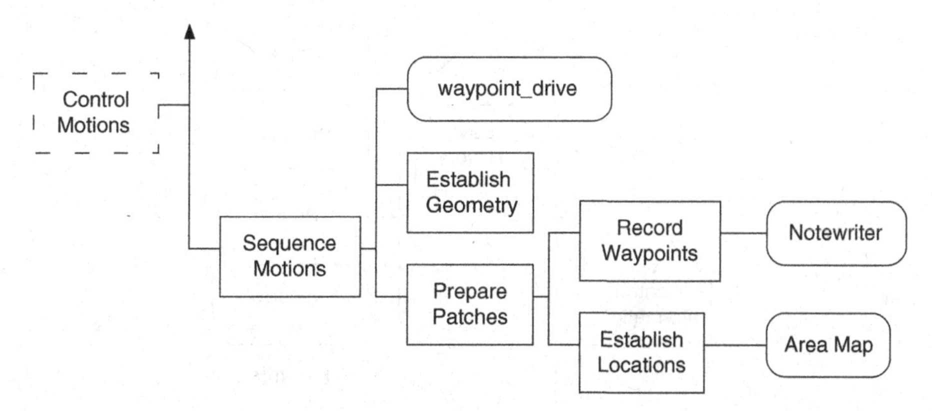


**Figure 6.25** Function diagram portion for *Control Motions—Sequence Motions*

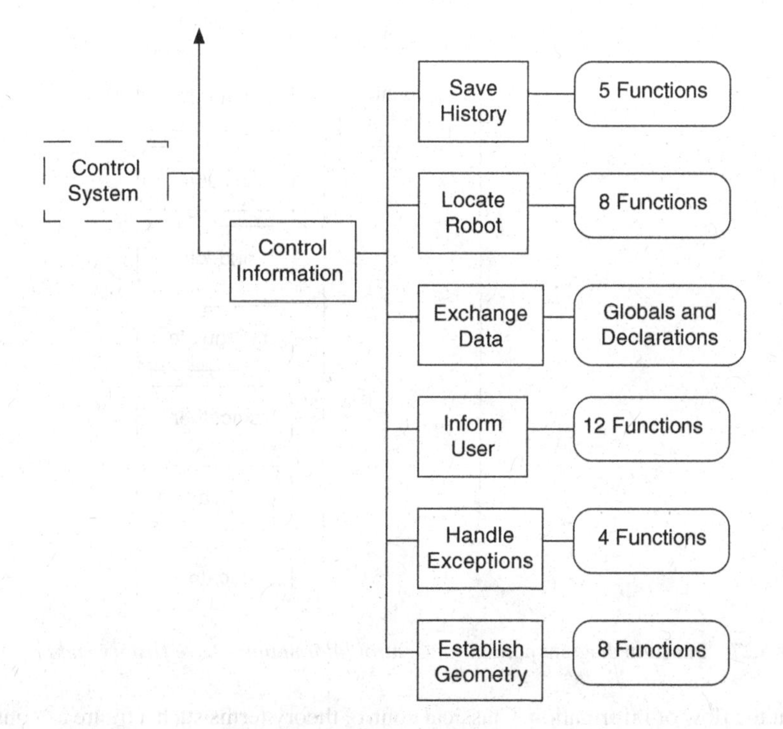


**Figure 6.26** Function diagram portion for *Control System—Control Information*

entire "history" can fit on one printed page in easily readable type. This length has proven to be useful for diagnosing behavior of the robot and the system. The width tends to be varied depending on the experiment performed, but generally, the following four stacks are represented: control variables, position values, distance values used in triangulation, and a list of flags which indicate any problems with receiving distance readings.

The major control variables comprise elements of feedback for line-following control, as shown in Figure 6.29. In this figure, the boxes represent numbers as named, and the arcs

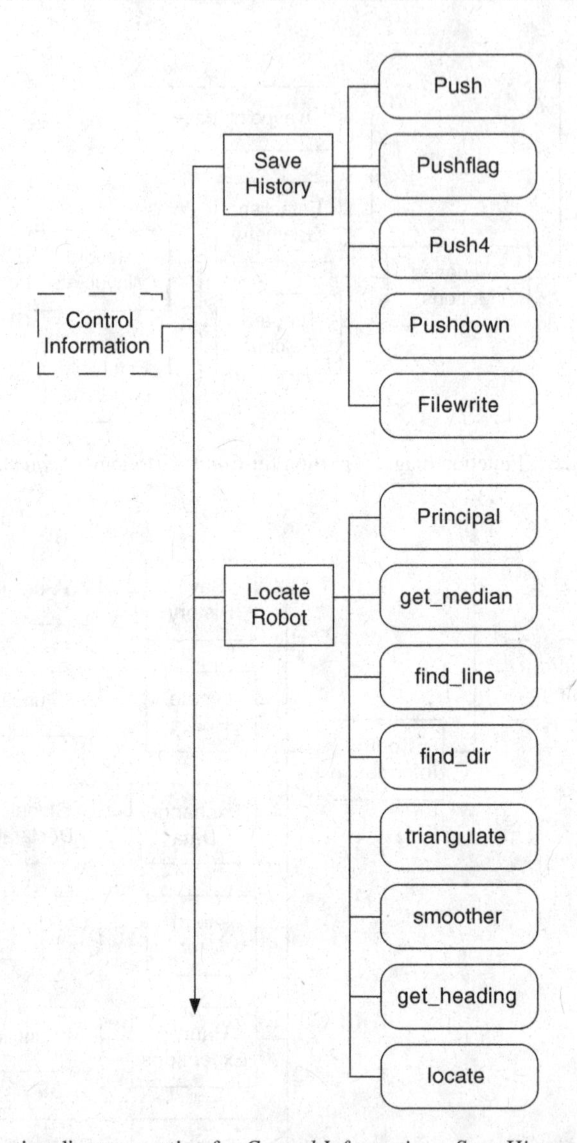


**Figure 6.27** Function diagram portion for *Control Information—Save History and Locate Robot*

represent the flow of information. Classical control theory terms such a figure a "control block diagram," and it details the inputs, outputs, and internal values used to regulate the lateral position of our robot with respect to the line that it is following. We employed the root-locus technique to set these values theoretically, as a starting point for experiments. These methods are well-known to controls engineers, but beyond our scope. We would recommend a comprehensive text[4] to describe and detail an intuitive set of methods for establishing stable controls for robotic devices. Every field robotics team should have, we think, a person dedicated to establishing the optimal values for the various control parameters, usually servo feedback gains. In this way every controlled task is accomplished through feedback of the errors encountered and then corrected. While control theory is beyond our scope, we will show how to establish reasonable values by experimental characterization and behavior of a system.

```
         CPhidgetEncoderHandle ENC = 0;         //Declare an Encoder handle for the 1047
12       CPhidgetAdvancedServoHandle SRV = 0; //Declare a Servo handle for the 1061
13       double const PI=3.141592654;                      //pi
14       int current_getch;                    //any character read from keyboard
15       int doloop = 1;                       //true, used for character input
16       float val[12]={0};                    //all the sensor readings from the encoders
17       static WINDOW *mainwnd;               //console window location pointer
18       static WINDOW *screen1;               //picture-in-picture loc pointer
19       static WINDOW *screen2;               //picture-in-picture loc pointer
20       static WINDOW *screen3;               //picture-in-picture loc pointer
21       static WINDOW *screen4;               //picture-in-picture loc pointer
22       int manual = 1, ret=0;                //general indicators for any function
23       const float SOS = 960;                //speed of sound in air, reciprocal
24       const float OFFSET=24;                //mike distance correction due to electronics
25       const int ZERO=101;                   //servo zero,switch side & board side
26       char value[10];                       //for keyboard inputs
27       char que[3] = "q\n";                  //quiting in extremis
28       double mov[8] = {0};                  //current servo commands
29       double vel=0;                         //special to drive functions
30       float old[12]={0}, deriv;             //part of data extraction from the encoders
31       float dist[12]={0}, avg[12]={0};//distance values, expect only 2 at a time
32       int flag[12]={0};                     //diagnostic tool for signal loss
33       SYSTEMTIME time;                      //for time-of-day
34       WORD millis;                          //in milliseconds
35       float X[13], Y[13], D[17];            //receiver locations tables
36       float xup=0, xum=0, yup=0, yum=0;//this is done to pass variables for triangulate
37       float xe=0, ye=0, ze=0, coord[6];     //our noble estimates, with deltas
38       int ca=0, cb=0;                       //user global receiver numbers, should not be equal
39       float da=0, db=0;                     //user global receiver distances
40       int pushdown(int column);             //function to store xe, ye, ze
41       float stack[61][6]={0};               //history locations
42       int stack_length = 60;                //height of stack for 1.5 seconds
43       float dex[61]={0},xbar;               //list for standard deviation
44       int point_list[100][5]={0};           //waypoints comprising "next" x, y and velocity
45       int point_count=0, n=0, cd=10;        //where we are in the list, global shut-down on n
46       double m, bi, xm, ym, mp, bp;                  //results of a line-find
47       double xi, yi, xp, yp;                         //results of an intersection
48       double arg1, arg2, arg3;                       //results of a circle-find for twisting
49       int push(int stack_num);                 //declare function for stuffing the distance stacks
50       int flag_stack[61][8];                         //diagnose loss of signal conditions
51       float dist_stack[61][8] = {0};        //distance stack, for now, may need 12
52       float smoothed_dist[61][8] = {0};    //smoothed stack, for now, may need 12
53       float stack4[61][4] = {0};                     //run stack, for now need 4 values
54       float xd, yd, tar, xs, ys;                     //waypoint destinations
55       float diff, error, d2line, delta, ex, ey;      //run values
56       FILE *workfile;                                //save up history
57       float find_line(int column);                   //declare outlier corrector
58       float find_avg(int column);                    //declare a simple average
59       float get_heading(float x1, float y1, float x2, float y2);//&c
60       float find_point(int column);                  //declare yet another best fitter
61       float stnd_dev(int n);                         //standard deviation finder
62       int pushflag(int column);                      //store any signal losses
```

**Figure 6.28** Portion of the program code: globals and declarations

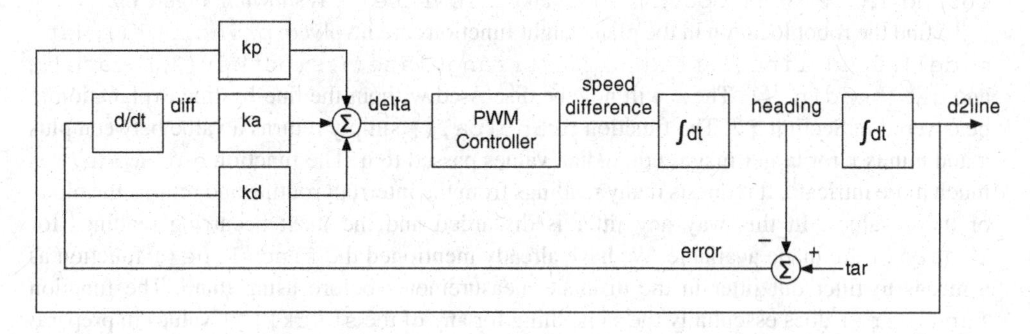


**Figure 6.29** Control block diagram for the field robot

The variable names shown in Figure 6.29 represent real physical quantities that are measurable and important for the system function *Follow Lines.* Starting from the right-hand side, the system output is the distance to the prescribed line between waypoints as measured by the `triangulate()` function, the value in `d2line`. It is this distance that controls the *Skid Steer* function and causes the robot to locate correctly on the field. This value is actually determined by a software function to be described in Section 7.2 called `dist2line()` which depends indirectly on `triangulate()`. The distance to the line naturally happens by heading the robot in a certain direction and waiting a short time. This waiting constitutes a time-integration and noted by the symbol "$\int dt$." In the same way, setting a speed difference between each wheel changes the heading by simply waiting a short time. The `error` in the heading and the target line is found by simply subtracting the two at every cycle. Finally, we find the rate of change ($d/dt$) of the heading to determine how fast we should respond to the `error`. We do this by finding the difference between the present `error` and the previous one, and call it `diff`. To control the robot, we multiply each of the three by a "gain" value, to be determined by theory and experiment, and add them all together to find the command to give the PWM controller. We called this value `delta`. When it is zero, the speeds are the same, as we will see a bit later, and equal to a value called `vel`. These variables all appear in the function `waypoint_drive()`. This software function will also be discussed with respect to a function block diagram in Section 7.3.

The values saved on `stack4[][]` by the function `push4()` are just the four discussed above: `diff`, `delta`, `error`, and `d2line`. In this way, one can track the performance of the *Skid Steer* control portion of `waypoint_drive()`. For example, one may find after a short run, that `filewrite()` has saved these values in a file called `work`. If the starting point of the robot is very different than the point required by the file read by `fileread()`, one may find that `delta` hits its limits for most of the cycles recorded as the robot attempts to make a sharp turn to the required starting point. To determine exactly what `filewrite()` will store, please refer to Section 7.4.

The next values saved by `filewrite()` are the `x`, `y` and heading, or `z`, values that were experienced by the robot as determined by the `locate()` function. These values are given in decimal inches, and were saved by the `pushdown()` function onto the `stack[][]`. One can see that the heading values are derived from changes in the `x` and `y` values, since this robot does not carry a compass or other independent heading instrument. Within the software, these three values are called `xe`, `ye`, and `ze`, to indicate that these are estimates only and not error-free values. The next two columns of values saved by `filewrite()` are the distance measurements themselves saved by `push()` onto the `dist_stack[][]` and used by `locate()`. Only two are produced by the `InputHandlerE()` function. The remaining columns display the flags set when the distance measurement fails to return values. In total, five functions are used to *Save History:* `push()`, `pushdown()`, `push4()`, `pushflag()`, and `filewrite()`, as shown in Figure 6.27.

To find the robot location in the plane, eight functions are involved: `principal()`, `get_mode()`, `find_line()`, `find_dir()`, `triangulate()`, `smoother()`, `locate()`, and `get_heading()`. These will now be discussed without the line-by-line explanation to be covered in Section 7.2. The function `principal()` simply returns a value between plus $\pi$ and minus $\pi$ for larger magnitude radian values passed to it. The function `get_mode()` is much more intricate. It requests many readings from the interrupt routine and returns the mode of those values. In this way any jitter is discarded and the most occurring readings for `locate()` are made available. We have already mentioned the `find_line()` function as a means to filter out jitter in the distance measurements before using them. The function `find_dir()` does essentially the same thing for any of the `stack[][]` values in preparation for finding the most reliable heading value for `ze`. The function `triangulate()`

solves for the triangles that satisfy the distance measurements, with respect to the computed distance between fixed pair of receivers. The receiver locations are saved by number in a predetermined list, as mentioned above. Since there are always two possible triangles, this function selects the result that is closest to the current `xe` and `ye` values. The `smoother()` function finds the best heading, or `ze` value given two sequential pairs of `xe` and `ye` values. The `locate()` function itself checks to see if the current `xe` and `ye` values are zero, and if so, asks the user for a set that may put `triangulate()` in range of a solution. If the robot is not moving, it requests a set of distance values based on `get_mode()`, otherwise it uses the current distance values computed by the `InputHandlerE()` function running in the background.

In order to *Inform User* of software progress, there are 12 functions to be called that influence the user screen. See Figure 6.30 for the function diagram. We have already

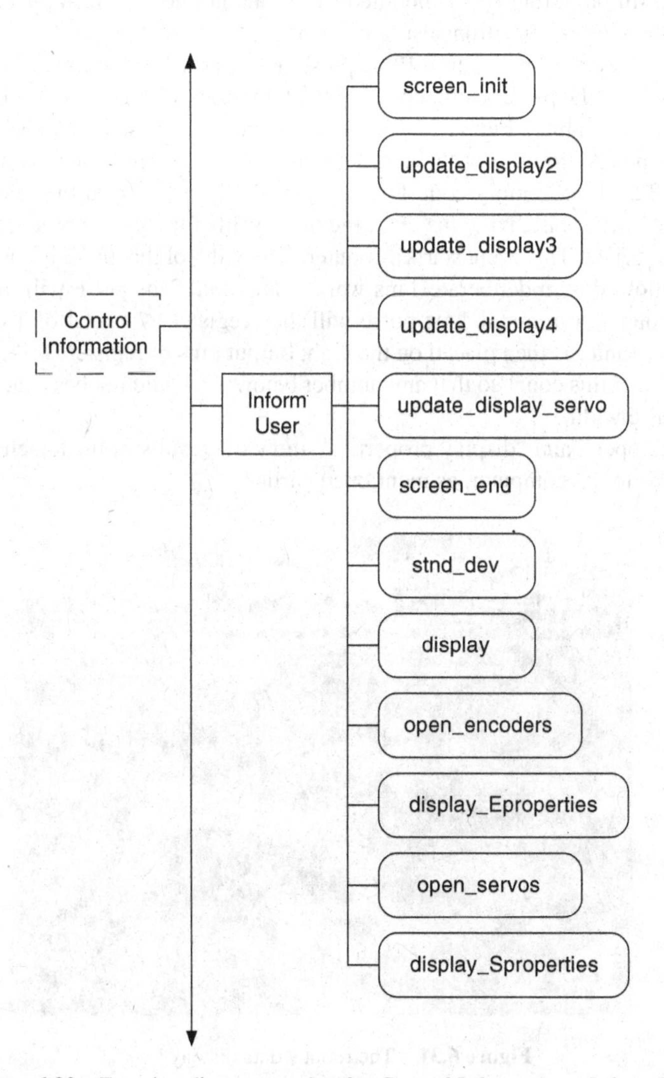


**Figure 6.30** Function diagram portion for *Control Information—Inform User*

mentioned a few of these, but the complete list is now given here: `screen_init()`, `update_display2()`, `update_display3()`, `update_display4()`, `screen_end()`,`update_display_servo()`,`stnd_dev()`,`display()`,`open_encoders()`, `display_Eproperties`, `open_servos()`, and `display_Sproperties()`. The `screen_init()` function prepares the user screen for use, while the `screen_end()` function closes it down. Both are `curses.h` functions. The four "update_display" functions place current values in separate windows on the screen for easy reference. The `stnd_dev()` function computes the standard deviation of any series of values loaded into the list `dex[]`. It is most useful for finding the optimum servo gains by giving analyses of the four control values mentioned above. The `display()` function stores a number of fixed parameters that convert the width of servo channel 3 into a binary representation for display on the robot to the user. It is part of the way one can obtain real-time information about which codes are running and their status through four robot-mounted LEDs and not need to scan the computer screen. At this time, there are no functions using its data.

The means for converting a servo PPM pulse into a four-bit number is in three steps: we need a way to sense the pulse width, count up the total time with a clock, and display this time as a digital byte (eight bits). Figure 5.32a shows a board to do these functions just left of center on the front panel. A close-up of the board and its circuit schematic are given in Figure 6.31 and Figure 6.32. The assembly code for the 16C54 chip is given in Figure 6.33. It is quite straightforward: after initializing the chip, the input waits for a pulse coming in from channel 3 of the main receiver. This event starts a counter. The width of that pulse is constantly checked every 4 μs, followed by a deliberate 1 ms wait. A maximal 2 ms pulse will then register as a total of 250 counts; a minimal 1 ms pulse will then register as a total of 1 count. The total number in the counter is then placed on the eight output pins of register B. We have used only the top four bits of this count so that any number below 16 would not be seen. This avoids any jitter in the pulse width.

Finally, the "open" and "display properties" show the results of the attachment process of the USB boards to the computer, as mentioned earlier.

**Figure 6.31** The robot's data display board

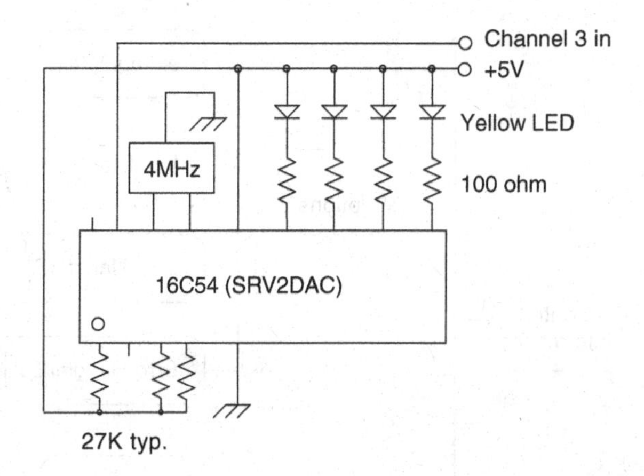


**Figure 6.32** Circuit schematic for the display board

```
          ;SRV2DAC.SRC
          DEVICE   PIC16C54,XT_OSC,WDT_OFF,PROTECT_OFF
          RESET    SETIO
OUT       EQU      RB          ;this code reads hobby servo
IN        EQU      RA.0        ;pulses and puts out the
SIGNAL    EQU      RA.1        ;width as 8 bits for a DAC
MODE      EQU      RA.2        ;mode = centered or zeroed
INV       EQU      RB.3        ;invert output
X         EQU      10H         ;counter for pulse width
Y         EQU      11H         ;scratch counter for min. pulse
SETIO     CLR      RA
          CLR      RB
          MOV      !RA,#00001111B    ;four inputs bits
          MOV      !RB,#00000000B    ;8 output bits
IDLE      MOV      RB,#0       ;default is no output
LOOK      JNB      RA.1,IDLE:if no pulses, wait here
          NOP
WAIT      JNB      RA.0,WAIT;if no pulse,wait here
          CALL     ONEMS       ;got a pulse, wait 1 ms
          NOP
          MOV      X,#0        ;clear width counter
INX       INC      X           ;bump count
          ;NOP                  ;may be needed if too fast
          JB       RA.0,INX   ;if pulse still there
DONE      MOV      RB,X        ;latch counted value
          JMP      LOOK        ;return to beginning

ONEMS     MOV      Y,#250      ;check this for speed
          ;SETB     RB.1
MS        NOP
          DECSZ    Y           ;count down
          JMP      MS          ;done?
          ;CLRB     RB.1
          RET                  ;yes
```

**Figure 6.33** The assembly code for converting a servo pulse width into a binary number

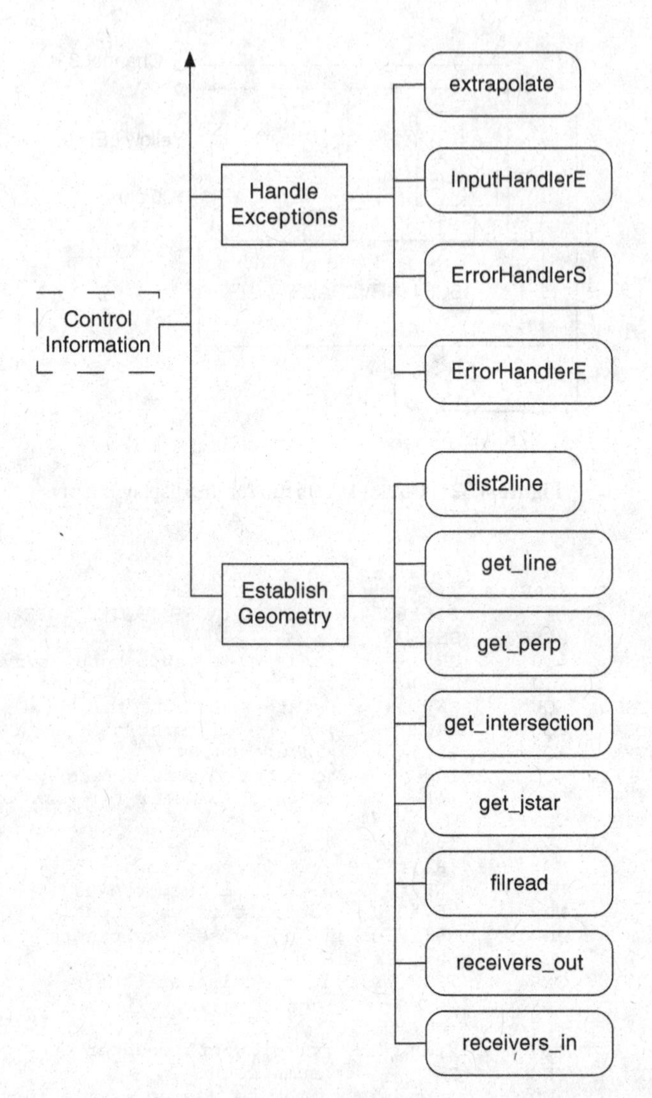


**Figure 6.34** Function diagram portion for *Control Information—Handle Exceptions* and *Establish Geometry*

In order to *Handle Exceptions,* there are four functions needed, as shown in Figure 6.34: `InputHandlerE()`, `ErrorHandlerS()`, `ErrorHandlerE()`, and `extrapolate()`. The first of these has already been discussed in detail in Section 5.4. It runs in the background responding to the hardware interrupts that accompany measurement of time-of-flight distances. The last of these has also been discussed at length in Section 5.4, and serves to keep the robot located even if the distance measurements fail for a brief period. The "error handler" codes report any problems connecting the USB boards to the computer. No other exceptions are considered by the software.

Finally, to *Establish Geometry,* there are eight functions: `dist2line()`, `get_line()`, `get_perp()`, `get_intersection()`, `get_jstar()`, `fileread()`, `receivers_out()`, and `receivers_in()`. The first of these, `dist2line()`, calculates the distance from a point to a given line. This function is used two ways: first, to find the position, `d2line`, of the robot relative to the line connecting two waypoints, and secondly, to find the distance to go, `d2go`, until the next waypoint is reached. Therefore, it is a key value used to control the robot position while between waypoints (Section 5.4), and it computes the end of the inner `while` loop in `waypoint_drive()`. The way in which `d2go` is computed deserves some detailed discussion here. Since the exact position of the robot in the vicinity of a target waypoint may not be accurate, and we do not wish the robot to spiral into such a point, we find the slope and intercept of the line connecting the waypoints, and then find the perpendicular to that line at the target waypoint, as shown in Figure 6.35. The robot then has a line to cross rather than a point to find in the plane. The robot may have a bit of lateral error and still complete the move to the next waypoint successfully. The distance to that line is `d2go`, and the control loop terminates when `d2go` is sufficiently small.

The function `get_intersection()` is used to find the point at which the robot should be when passing between receivers and cannot `triangulate()`. This construction point, `xi,yi` in Figure 6.36, is compared with its current position in the `get_jstar()` function so that an estimate of the total distance that the robot will be "in the dark" can be found. The value `jstar` represents the number of `while` loop cycles need to complete the move without being able to `locate()`.

The last three functions establish the geometry needed to represent a lawn patch, or the receiver locations, respectively. Chapter 7 will detail all of the software briefly described above on a line-by-line basis.

Before we leave the mechatronics part of our book, it would be useful to portray the overall system as a set of components previously described functionally, but not explicitly put into a diagram. A view of the System Components is given in Figure 6.37, with the caveat that components are *not* functions, but we do need real artifacts to *Mow Lawn.* The arrangement shown includes a new part, taken for granted in the connection of peripherals to a netbook: a USB hub to connect the two Phidgets boards in the USB translator to the PC. The Receivers would be located in the lawn area, much as shown in Figure 4.2, and wired to the circuits of the USB translator as shown in Figure 5.15. The Robot is free to roam within the joint 20-m

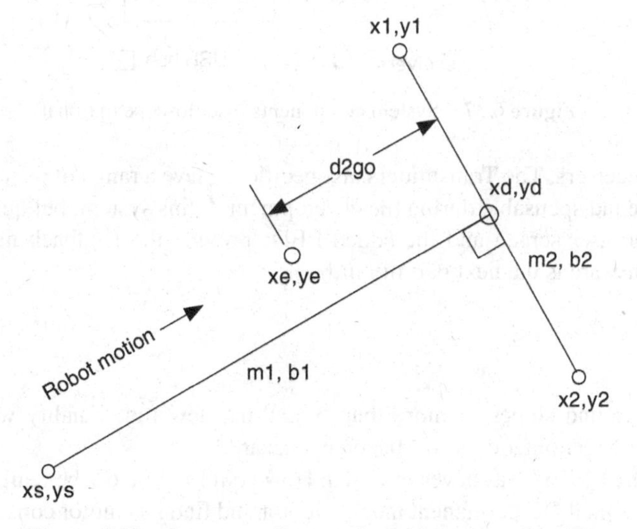


**Figure 6.35** Geometry of computing a distance to Go, d2go

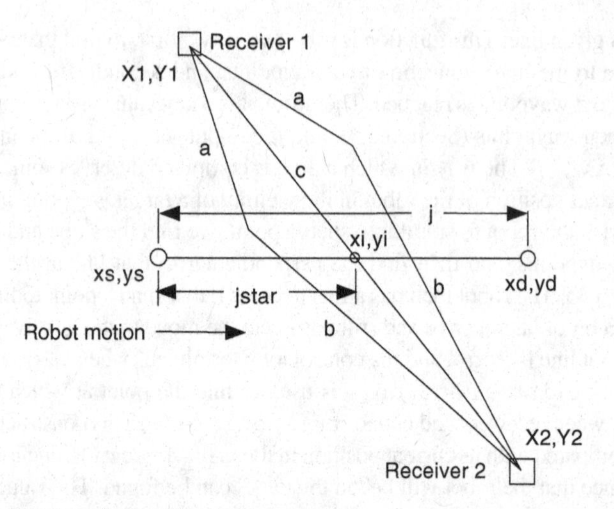


**Figure 6.36** Geometry of computing a crossing point between receivers

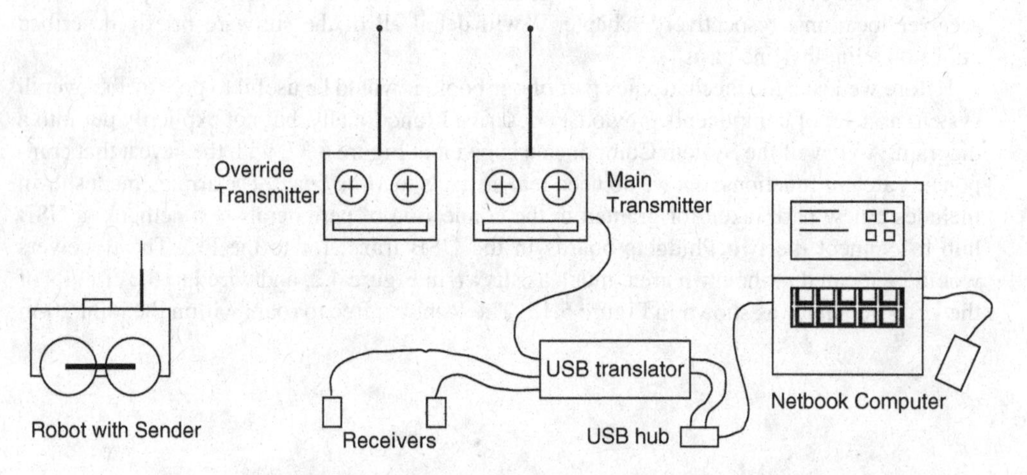


**Figure 6.37** System components (oscilloscope optional)

range of the Receivers. The Transmitters, as specificed, have a range of over 500 m. We found an oscilloscope indispensable during the development of this system, but quite optional when completed. The user screen and the added LEDs provide the feedback needed to operate. Operational software is the next part of our book.

## Problems

6.1 Consider ground slopes of more than 5%. What new functionality would be needed? When does an automatic mower become a hazard?

6.2 Suppose the LM567 was never invented. How could Figure 6.7 be realized?

6.3 Calibrate a small DC permanent magnet motor and find its "motor constant," that is, what value k satisfies both $V = k\,\omega$ and $\tau = k\,I$ ?

6.4 Suppose our robot "bounced around" on uneven turf. What could we do to *Mow Lawn* successfully?
6.5 Please rearrange the right-hand portion of Figure 6.2 for a "better" decomposition of functions, with the reasons for these alternatives.
6.6 Explain the reason for the matching of "`case 97:`" with the "`a`" key.
6.7 Please give examples of functions that cannot be accessed by a single keystroke from `main()` and explain why.
6.8 Explain the apparent disparity between the "ground" lines of Figure 5.9 with those of Figure 6.5, and show how the sound sending board is actually wired to the *Power Subsystems* circuit.
6.9 Explain Figure 6.8 in detail.
6.10 Where in the software is a "log" kept of the recent x and y positions?
6.11 Consider a new "backup maneuver" in case of a robot stall condition. What would comprise its function diagram? Please sketch its software block diagram.
6.12 Investigate the best types of filters for Gaussian noise in a controller and describe them.
6.13 Find the relationship between spacer bar hole spacing (Figure 6.20), chain pitch, and sprocket diameter to maintain an ideal chain tightness.
6.14 Please describe in functional terms the differences between a three-wheeled caster-type geometry, and a skid-steered design. How would these be manifest in terms of the number of functions and parts?
6.15 Give reasons for the skid-steer selection versus omnidirectional wheels.
6.16 Suppose `d2line` had almost no affect at all on `delta`. How would you change `kp`? When would you know to stop changing it?
6.17 Consider `update_display_servo()`. Rate its usefulness to the user.
6.18 Why is `update_display4()` useful?
6.19 Consider Figure 6.4 and suggest how to eliminate almost all of the declarations. Hint: refer to the Software Concordance of Appendix E.
6.20 Describe in your own words the function of the integrals shown in Figure 6.29. What would happen to the robot if they were not compensated for?
6.21 The robot heading is not shown anywhere in the code, yet it needs to be compensated. What is computed instead of the robot heading?
6.22 Why should the *mode* be used rather than the *average* when removing jitter from distance readings?
6.23 What would likely be the motion of the robot if only `xd, yd` were the end point rather than the value `d2go`?
6.24 What would likely be the motion of the robot if `jstar` were never calculated?
6.25 Please consider the seven connecting cables in Figure 6.37. What signals and power are carried on each of these?

## Notes

1. Bergsman, P. (1994) *Controlling the World with Your PC,* Newness Press.
2. IEEE 1394 (1994–present) Firewire standard, en.wikipedia.org/wiki/IEEE_1394.
3. Fowler, T.C. (1990) *Value Analysis in Design,* Van Nostrand Reinhold.
4. Phillips, C.L. and Harbor, R. (1988) *Feedback Control Systems*, Prentice Hall.

# 7

# Software Functions

At this point, the reader should have a complete picture of the "how's" and "why's" of both the hardware and software of the *Mow Lawn* system. The software taken alone need not have function diagrams, since they already appear and form part of the system already functionally decomposed. To be most useful, however, the software needs to be described on a more detailed basis. The user need not "reverse engineer" our code, a task that may be most difficult and frustrating, especially if he/she is not practiced in the C-language. Tabs were removed for publishing, but may be freely added as needed for readability.

This chapter can be used as an index to specific points in the code, or read as a whole to understand every decision made. Each line is commented for reference. Certain functions have been described in detail earlier, and they are repeated here for context and clarity. After the preliminary `#includes` and `globals`, this chapter is further subdivided into logical sections:

Displays: to place needed information to the user screen.
Field data and triangulation: geometric locating functions.
Operation: the calls that make the robot move and stop.
History and diagnostics: the immediate past used for analysis.

The following comments appear at the head of the code. It gives the last date of instantiation and a few words about its recent history. Unlike the robot hardware, built according to Function Logic, each code element represents a single function that connects to selected others via the PlanD Software Concordance given in Appendix E.

```
//PlanD, built on an old ifKit example, 4 August 13
//This version reads the encoders and does its own USB
//it also uses feedback from x,y to create z orientations
//new functions for emergency back-ups, and accel limits added
```

*Practical Field Robotics: A Systems Approach*, First Edition. Robert H. Sturges, Jr.

Companion Website: www.wiley.com/go/sturges

The following lines declare the library files needed and their reasons for inclusion.

```
#include "stdafx.h" //needed for all c programs
#include "phidget21.h" //needed to access the USB device boards
#include "curses.h" //needed for fancy screen manipulation
#include < Windows.h>//needed for Sleep function
#include < math.h>//needed for 'atof' & trig functions
#include < stdio.h>//a thneed is a something that everyone needs
```

The following lines declare global variables, constants, and functions and their purposes.

```
CPhidgetEncoderHandle ENC = 0; //Declare an Encoder handle for the
1047
CPhidgetAdvancedServoHandle SRV = 0; //Declare a Servo handle for the
1061
double const PI = 3.141592654; //pi
int current_getch; //any character read from keyboard
int doloop = 1; //true, used for character input
float val[12] = {0}; //all the sensor readings from the encoders
static WINDOW *mainwnd; //console window location pointer
static WINDOW *screen1; //picture-in-picture loc pointer
static WINDOW *screen2; //picture-in-picture loc pointer
static WINDOW *screen3; //picture-in-picture loc pointer
static WINDOW *screen4; //picture-in-picture loc pointer
int manual = 1, ret = 0; //general indicators for any function
const float SOS = 960; //speed of sound in air, reciprocal
const float OFFSET = 24; //mike distance correction due to
electronics
const int ZERO = 101; //servo zero,switch side & board side
char value[10]; //for keyboard inputs
char que[3] = "q\n"; //quiting in extremis
double mov[8] = {0}; //current servo commands
double vel = 0; //special to drive functions
float old[12] = {0}, deriv; //part of data extraction from the
encoders
float dist[12] = {0}, avg[12] = {0};//distance values, expect only 2
at a time
int flag[12] = {0}; //diagnostic tool for signal loss
SYSTEMTIME time; //for time-of-day
WORD millis; //in milliseconds
float X[13], Y[13], D[17]; //receiver locations tables
float xup = 0, xum = 0, yup = 0, yum = 0;//this is done to pass
variables for triangulate
float xe = 0, ye = 0, ze = 0, coord[6]; //our noble estimates, with
deltas
int ca = 0, cb = 0; //user global receiver numbers, should not be
equal
float da = 0, db = 0; //user global receiver distances
int pushdown(int column); //function to store xe, ye, ze
float stack[61][6] = {0}; //history locations
int stack_length = 60; //height of stack for 1.5 seconds
float dex[61] = {0}, xbar; //list for standard deviation
int point_list[100][5] = {0}; //waypoints comprising "next" x, y and
velocity
```

```
int point_count = 0, n = 0, cd = 10; //where we are in the list,
global shut-down on n
double m, bi, xm, ym, mp, bp; //results of a line-find
double xi, yi, xp, yp; //results of an intersection
double arg1, arg2, arg3; //results of a circle-find for twisting
int push(int stack_num); //declare function for stuffing the distance
stacks
int flag_stack[121][8]; //diagnose loss of signal conditions
float dist_stack[121][8] = {0}; //distance stack, for now, may need
12
float smoothed_dist[121][8] = {0}; //smoothed stack, for now, may
need 12
float stack4[121][4] = {0}; //run stack, for now need 4 values
float xd, yd, tar, xs, ys; //waypoint destinations
float diff, error, d2line, delta, ex, ey; //run values
FILE *workfile; //save up history
float find_line(int column); //declare outlier corrector
float find_avg(int column); //declare a simple average
float get_heading(float x1, float y1, float x2, float y2);//&c
float find_point(int column); //declare yet another best fitter
float stnd_dev(int n); //standard deviation finder
int pushflag(int column); //store any signal losses
```

## 7.1 Displays: To Place Needed Information to the User Screen

The following functions deal with the user screen in displaying two different kinds of information. One kind appears line-by-line in order of occurrence. Most of these screen-display instructions are secondary to the function they occur in. The second kind appears as lists in boxed windows on the screen, always in the same places. They take precedence over the line-by-line information displays. The following functions in this category are called by others as indicated in the Software Concordance included in Appendix E.

This function is called by `main()` and uses the `curses.h` library to start the screen operations.

```
void screen_init(void) { //startup the console screen
mainwnd = initscr(); //cls and move cursor to 0,0
cbreak(); //be ready to stop the window, too
nodelay(mainwnd, TRUE); //immediately use the getch()
refresh(); //show off all the text in the little window
wrefresh(mainwnd); //show off the main console window
screen1 = newwin(13, 27, 1, 67); //define the sensor display box
box(screen1, ACS_VLINE, ACS_HLINE);//draw a box around it
screen2 = newwin(11, 26, 1, 96); //define the sensor display box
box(screen2, ACS_VLINE, ACS_HLINE);//y-size, x-size, y-loc, x-loc
screen3 = newwin(17, 24, 15, 100); //define the sensor display box
box(screen3, ACS_VLINE, ACS_HLINE);//y-size, x-size, y-loc, x-loc
screen4 = newwin(17, 32, 15, 67); //define the path display box
box(screen4, ACS_VLINE, ACS_HLINE);//y-size, x-size, y-loc, x-loc
}
```

This function updates the display of the commands currently given to the two motor/wheel servos and the third channel used for display purposes. It needs special print commands since it is working through the `curses.h` library. "`_servo`" identifies the first box on the screen.

```
// function to update the screen with current servo moves
static void update_display_servo(double *mov) {
curs_set(0);
mvwprintw(screen1,1,3,"Servo %d moved to %4 g",0,mov[0]);
mvwprintw(screen1,3,3,"Servo %d moved to %4 g",1,mov[1]);
mvwprintw(screen1,5,3,"Servo %d moved to %4 g",2,mov[2]);
wrefresh(screen1);
}
```

This function updates the display of the distances as found by two receivers, taking their ID numbers modulo 2. It needs special print commands since it is working through the `curses.h` library. "`_display2`" identifies the second box on the screen.

```
static void update_display2(float *avg) { //for only 2 mikes now
curs_set(0); //init the cursor
mvwprintw(screen2,1,1,"----VALUES in INCHES----"); //fancy start
mvwprintw(screen2,3,3,"Mike %d measures %d ",ca,(int)da);//show what
is found
mvwprintw(screen2,4,3,"Mike %d measures %d ",cb,(int)db);//show what
is found
mvwprintw(screen2,6,6,"PRESS q TO END"); //reminder to user
mvwprintw(screen2,8,1,"----THAT'S IT TODAY!----"); //fancy end
wrefresh(screen2); //put it up to see
}
```

This function updates the display of the estimates of the x and y distances (xe and ye) as found by triangulating the two receivers, and using their ID numbers with respect to the table "`receivers_out`" or "`receivers_in`." It also displays the heading value, ze, as found by finding the angle between the differences of a number of xe and ye values. The "`smoother()`" function does this computation. The window displays 12 of these values from the stack, so the user can see the recent history of about a 1 s time period. Again, it needs special print commands since it is working through the `curses.h` library. "`_display3`" identifies the third box on the screen.

```
static void update_display3() { //for 12 stack values
int i;
curs_set(0); //init the cursor
mvwprintw(screen3,1,1," xe ye ze"); //fancy start
for(i = 0;i < 12;i++)
mvwprintw(screen3,3 + i,1,"%4d %4d %4.2f",\
(int)stack[i][0],(int)stack[i][1],stack[i][2]);
wrefresh(screen3); //put it up to see
}
```

This function updates the display of the current control variables as found by the servo control law. The window displays 12 of these values from the stack, so the user can see the recent history of about a 1 s time period. Again, it needs special print commands since it is working through the `curses.h` library. "`_display4`" identifies the third box on the screen.

```
static void update_display4() { //for 4 run values
int i;
curs_set(0); //init the cursor
mvwprintw(screen4,1,1," diff delta error d2L"); //fancy start
for(i = 0;i < 12;i++)
mvwprintw(screen4,3 + i,1," %5.2f %5.2f %5.2f %d",\
stack4[i][0],stack4[i][1],stack4[i][2],(int)stack4[i][3]);
wrefresh(screen4); //put it up to see
}
```

This function formally closes the screens setup in the `screen_init()` function using the curses.h library.

```
// function to close the screens
void screen_end(void) {
endwin(); //special function in curses.h
}
```

## 7.2 Field Data and Triangulation: Geometric Locating Functions

The following functions deal with obtaining and using data from the receivers and setting up the geometry of the field itself. Most important to any mobile robot application is the ability to self-locate in what we have termed "the x,y problem." The initialization functions come first, followed by the locate functions called every time the control needs updates.

This function loads global variables with the geometric locations of the receivers in the plane of the robot in the outdoor field. Actual measured values are in inches. Ones indicate no values have been assigned at this time.

```
void receivers_out() {
X[0] = 1432; Y[0] = 591; //across from abelias
X[1] = 1144; Y[1] = 518; //trash can corner
//X[2] = 801; Y[2] = 398; //east deck corner, test location
//X[3] = 498; Y[3] = 408; //west deck corner, test location
X[2] = 1004; Y[2] = 50; //east 2A post
X[3] = 723; Y[3] = 54; //west 3A post
X[4] = 1; Y[4] = 1; //south-west house corner
X[5] = 1; Y[5] = 1;
X[6] = 1; Y[6] = 1;
X[7] = 1; Y[7] = 1;
X[8] = 1; Y[8] = 1;
X[9] = 1; Y[9] = 1;
X[10] = 1; Y[10] = 1;
X[11] = 1; Y[11] = 1;
printw(" OUTSIDE! "); //remind the user
}
```

This function loads global variables with the geometric locations of the receivers in the plane of the robot in the indoor laboratory field. Actual measured values are in inches. Ones indicate no values have been assigned at this time.

```
void receivers_in() {
X[0] = 1; Y[0] = 1;
X[1] = 165; Y[1] = 146; //end of bookshelves
```

```
X[2] = 196; Y[2] = 142; //teak table leg
X[3] = 26; Y[3] = 243; //drill press table leg
X[4] = 1; Y[4] = 1;
X[5] = 1; Y[5] = 1;
X[6] = 1; Y[6] = 1;
X[7] = 1; Y[7] = 1;
X[8] = 1; Y[8] = 1;
X[9] = 1; Y[9] = 1;
X[10] = 1; Y[10] = 1;
X[11] = 1; Y[11] = 1;
printw(" INSIDE! ");
}
```

This function obtains the raw data from the USB translator box described in detail in Section 5.4 and Figure 3.5. Please refer to that text when reviewing each line of the code here. As a reminder, this function is called on an interrupt basis, that is, only when there is a change of state of the sync signal, which occurs every 80 ms. Since the sync signal is only a few microseconds long, and may be missed by the software, it is used to trigger a single-stage counter, as shown in Figure 3.1. The edges of this counter change whenever a sync signal arrives. The levels of this counter output essentially divide the sync signals by two, but are not used. During the execution of this function, two distance values are received and interpreted. Two distances are sufficient for triangulation.

```
int __stdcall InputHandlerE(CPhidgetEncoderHandle ENC, void *userptr,
\
int Index, int State)
{ //This is an encoder board event-triggered call-back function on
sync
int iposition; //get only 2 positions, ca & cb
CPhidgetEncoder_getIndexPosition(ENC,ca,&iposition);//get the true
val[ca] = iposition-old[ca]; //position based on the index, which we
set
if(val[ca]==0) val[ca] = old[ca]; //tied to USB-translator hardware
else old[ca] = (float)iposition; //every 80 ms. Save the prior value
dist[ca] = val[ca]*100/SOS-OFFSET; //convert to inches
CPhidgetEncoder_getIndexPosition(ENC,cb,&iposition);//get the true
val[cb] = iposition-old[cb]; //position based on the index, which we
set
if(val[cb]==0) val[cb] = old[cb]; //check out section && notes
else old[cb] = (float)iposition; //every 80 ms. Save the prior value
dist[cb] = val[cb]*100/SOS-OFFSET; //convert to inches
if(dist[ca] < 0 || dist[ca] > 900) { //check range
Beep(200,10); manual = 0; //if out, warn & set flag
flag[ca] = 1; dist[ca] = smoothed_dist[0][ca];} //extrapolate
dist[ca]
else {flag[ca] = 0; manual = 1;} //reset flag
find_line(ca); avg[ca] = (float)bi; //pass line through history
da = avg[ca]; push(ca); pushflag(ca); //post the avg distance da
if(dist[cb] < 0 || dist[cb] > 900) {
Beep(200,10); manual = 0; //sounds like a click
flag[cb] = 1; dist[cb] = smoothed_dist[0][cb];} //extrapolate
dist[cb]
else {flag[cb] = 0; manual = 1;} //reset flag
find_line(cb); avg[cb] = (float)bi; //and save the predictions
db = avg[cb]; push(cb); pushflag(cb); //post the avg distance db
update_display2(avg); //show off distance data
return 0;
}
```

This function computes the distance from a point (x3, y3) to the line described by points x1, y1 and x2, y2. It returns a signed value so we can determine which side of the line the point lies. This is a standard math function for the robot code. It is used to find the error `d2line` in location of the robot from the line between two waypoints. It is also used to find the distance from the current x, y position to the end of the move toward the next waypoint. This line, in the latter case, is perpendicular to the line connecting the waypoints and runs through the destination waypoint. his distance is `d2go`.

```
double dist2line(float x1, float y1, float x2, float y2, float x3,
float y3) {
return (((x2-x1)*(y1-y3)-(x1-x3)*(y2-y1))/ //signed value, + = right
side
sqrt((double)((x2-x1)*(x2-x1) + (y2-y1)*(y2-y1))));
//x1 and x2 are ON the line, x3 is a distance from it
}
```

This simple function takes a formal parameter x that corresponds to a heading angle in radians. If the value is not a "principal value" between plus and minus $\pi$, it makes the correction. The value x is assumed to be no greater in magnitude than $2\pi$ and is not checked.

```
float principal(float x) { //adjust angle to principal value
if(x > PI) x = x-2*(float)PI; //that is, between +/- PI
if(x < -PI) x = x + 2*(float)PI;
return x;
}
```

This function is used to convert the coordinates of two points (formal parameters) on a line to the slope and intercept of that line. It anticipates that slopes can go towards infinity, so it substitutes a large value in its place instead of causing an overflow. This value is sufficiently large to avoid positioning errors. It returns the slope as the global variable m, and the intercepts as the global variable `bi`. In addition, it calculates the perpendicular bisector of the line between the two given points and returns those values as `mp` and `bp`, respectively. This function is called by `get_jstar` so that an extrapolation can be made in case `triangulate()` fails.

```
int get_line(float x1, float y1, float x2, float y2) {//from two
points on line
if((int)x1==(int)x2)m = 1000; //infinite slope!
else m = (double)(y1-y2)/(double)(x1-x2); //get slope
bi = (double)(y2-x2*m); //get intercept
if((int)m==0) mp = 1000; //zero slope!
else mp = -1/m; //get the normal slope of the perp bisector
xm = (x1 + x2)/2; ym = (y1 + y2)/2; //get midpoint of the original
line
bp = (double)(ym-xm*mp); //get b of the perp. bisector
return 0;
}
```

This function is similar to part of `get_line()`. It computes the perpendicular line passing through the second of the point pairs, given as formal parameters `x1`, `y1` and `x2`, `y2`. It anticipates that slopes can go towards infinity, so it substitutes a large value in its place instead of causing an overflow. This value is sufficiently large to avoid positioning errors. It returns the slope as the global variable `mp`, and the intercepts as the global variable `bp`. This function is called by `waypoint_drive()` so that the distance to the end of a waypoint line segment, `d2go`, can be calculated.

```
int get_perp(float x1, float y1, float x2, float y2) {//from two
points on line
if((int)x1==(int)x2)m = 1000; //infinite slope!
else m = (double)(y1-y2)/(double)(x1-x2); //get slope
bi = (double)(y2-x2*m); //get intercept
if((int)m==0) mp = 1000; //zero slope!
else mp = -1/m; //get the normal slope of the perp line
bp = (double)(y2-x2*mp); //get b of the perp. through (x2,y2)
return 0;
}
```

This function is used to find the intersection of two lines given by their slopes and intercepts as formal parameters. It anticipates that slopes can go towards infinity, so it substitutes a large value in its place instead of causing an overflow. This value is sufficiently large to avoid positioning errors. It returns the coordinates of the intersection point as xi and yi as global variables. This function is called by `get_jstar()` so that an extrapolation can be made in case `triangulate()` fails.

```
int get_intersection(double m1, double b1, double m2, double b2) {
double epsilon; //for infinite cases
if((int)m1==(int)m2) epsilon = .001; //parallel lines!
else epsilon = 0; //only a tiny change
xi = (b2-b1)/(m1-m2 + epsilon); //solve for intersection by adding 2
lines
yi = m1 * xi + b1; //and substituting
return 0;
}
```

This function seeks the statistical mode of a pair of values from a series of values found by the interrupt routine `InputHandlerE`. This mode represents the best value of distance that can be measured given that there may be jitter in the measurement. The routine works by taking readings of the distances every 80ms so that data repetitions will not occur. For each of two distances it collects a list of values in "raw" form, that is, before being converted into inches. If a reading matches a previous one, a counter (`list[][]`) is incremented to build up a histogram. If a sufficient number of matches are found (20), the routine ends. It finds the peak number of occurrences and retrieves that pair as the modes. It converts the mode values to inches and returns them in `avg[0]` and `avg[1]`. It ignores data dropouts and zero values. This routine is called by `locate()` only if the robot is not moving. It is used to help locate the receiver stations and any other points of interest in the work space of the robot.

```
int get_mode() { //populate a table of non-zero values and their
frequencies
int list[10][8] = {0}; //guess at the table length, report it later
int i, j, match, mode = 0, peak; //useful variables
do{ Sleep(80); val[0] = val[ca]; val[1] = val[cb]; // get 2 values
for(j = 0;j < 2;j++) { //do for each encoder
match = 0; //no match assumed
if(val[j]! = 0) { //if zero, ignore it
for(i = 0;i < 10;i++) { //search the list
if(val[j]==list[i][j]) {//if matched, bump the count
list[i][j + 4]++; mode++; match = 1; break;
} //and pick up the next of 4 vals
}
if(match==0) { //we end up here if no match
for(i = 0;i < 10;i++) { //search the list
if(list[i][j]==0){ //found an open slot
```

```
list[i][j] = (int)val[j]; list[i][j + 4]++; break;
} //bump the count, yada, yada
}
}
} //the list is now populated for one pass
}} while(mode < 20); //the list now has many passes
for(j = 0;j < 2;j++) { //do for each encoder
peak = 0; match = 0; //clear cursors
for(i = 0;i < 10;i++) { //search the list for the highest frequency
if(list[i][j + 4] > peak) { //found a bigger one
peak = list[i][j + 4]; match = i; //save the value and the index
}
}
val[j] = (float)list[match][j]; //most popular val in that column
} //not yet converted to inches
clear(); //do this when you have something to show
for(i = 0;i < 10;i++) { printw("\n"); //show the list
for(j = 0;j < 8;j++) printw(" %d",list[i][j]);}
printw("\n %d %d \n",(int)val[0],(int)val[1]);
for(j = 0;j < 2;j++) avg[j] = (val[j]*100/SOS)-OFFSET; //convert to
inches
printw("\n %d %d \n",(int)avg[0],(int)avg[1]);
screen_init();
update_display2(val); //second window over
return 0;
}
```

This function computes smoothed-out values for the raw distance values computed in `InputHandlerE()`. It only uses one column of values in the distance stack as specified by the formal parameter `column`. The method that it employs is to first make up a sequential list of values for an x-axis comprising 10 integers. It then sums those integers. This is not a constant since the user can choose to select more or less than 10 previous values. It then accesses the first 10 values in the distance stack and sums those values. It then computes the product of the x-values and corresponding stack values. It then computes the square of each of the stack values. With this information, it computes the slope of the best fit RMS line passing through the `distance_stack[][]` values. It similarly computes the y-intercept of that line. It returns the slope in m and the intercept in bi, both of which are globals intended for such purposes. In this way, this function averages out the jitter that may be present in the received distance values and preserves any trend that may be forming over the last 10 `locate()` cycles. Using the y-intercept as the representative value for the distance anticipates any upward or downward trend. This function is called twice by the `InputHandlerE` on interrupt.

```
float find_line(int column) { //best fit RMS line to points in
dist_stack[10][c]
int i, n = 10; //our durable indexes
float xlist[10]= {1,2,3,4,5,6,7,8,9,10};//the stack has sequential
values
float sx = 0, sy = 0, sxy = 0, sx2 = 0; //needed for RMS calculation
for(i = 0;i < n;i++) { //fill sums with n values
sx = sx + xlist[i]; //a constant
sy = sy + dist_stack[i][column]; //column = the mike number: 0, 1
sxy = sxy + xlist[i]*dist_stack[i][column];//the big player
sx2 = sx2 + xlist[i]*xlist[i]; //sum of x-squares, a constant
}
m = (float)((sxy - sx*sy/n)/(sx2-sx*sx/n)); // formula for slope
bi = (float)(sy/n - m*sx/n); // formula for intercept
return 0; //the intercept is our distance
}
```

This function computes smoothed-out values for the raw distance values computed in `InputHandlerE()`. It only uses one column of values in the distance stack as specified by the formal parameter `column`. The method that it employs is to first make up a sequential list of values for the previous 10 distance measurements. It then computes the standard deviation of those values. The user can choose to select more or less than 10 previous values. It then accesses the first 10 values in the distance stack and compares them with a value 2 standard deviations above and below the mean value, `xbar`, which is computed by `stnd_dev[]`. With this information, it substitutes the average value for the distance measurements that are "out of bounds." In this way, this function averages out the outlier values that may be present in the received distance values, but ignores any trend that may be forming over the last 10 `locate()` cycles. This function is no longer used by the program since outliers are now too rare to be considered separately. In previous versions of the receivers, it was useful.

```
float find_point(int column){
int i; float sd; //index and stnd deviation
for(i = 0;i < 10;i++) dex[i] = dist_stack[i][column]; //fill the dex
stack
sd = stnd_dev(10); //find the stats on it
if(dist_stack[0][column] > (xbar + 2.*sd) || //if the point is too
high
dist_stack[0][column] < (xbar-2.*sd)) //if the point is too low
dist_stack[0][column] = xbar; //substitute the average value
return 0;
}
```

This function computes smoothed-out values for the raw `xe`, `ye`, or `ze` values computed in `locate()`. It only uses one column of values in the stack as specified by the formal parameter `column`. The method that it employs is to first make up a sequential list of values for an x-axis comprising 10 integers. It then sums those integers. This is not a constant since the user can choose to select more or less than 10 previous values. It then accesses the first 10 values in the stack and sums those values. It then computes the product of the x-values and corresponding stack values. It then computes the square of each of the stack values. With this information, it computes the slope of the best fit RMS line passing through the `stack[][]` values. It similarly computes the y-intercept of that line. It returns the slope in `m` and the intercept in `bi`, both of which are globals intended for such purposes. In this way, this function averages out the jitter that may be present in the received distance values and preserves any trend that may be forming over the last 10 `locate()` cycles. Using the y-intercept as the representative value for the distance anticipates any upward or downward trend. This function also returns a y-value that is historically 10 cycles in the past with respect to the smoothed-out value reported in `bi`. This function is no longer used by the program since jitter and outliers are now too rare to be considered separately. In previous versions of the code that smoothed out the `ze` values, it was useful. It is called by `get_ze()`.

```
float find_dir(int column) { //best fit RMS line to points in
stack[10][c]
int i, n = 10; //our durable indexes
float xlist[10] = {0,1,2,3,4,5,6,7,8,9};//the stack has sequential
values
float sx = 0, sy = 0, sxy = 0, sx2 = 0, y; //needed for RMS
calculation
for(i = 0;i < n;i++) { //fill sums with n values
sx = sx + xlist[i]; //a constant
```

```
sy = sy + stack[i][column];//column = the mike number: 0, 1, 2, ...
sxy = sxy + xlist[i]*stack[i][column];//the big player
sx2 = sx2 + xlist[i]*xlist[i]; //sum of x-squares, a constant
}
m = (float)((sxy - sx*sy/n)/(sx2-sx*sx/n)); // formula for slope
bi = (float)(sy/n - m*sx/n); // formula for intercept
y = (float)(m*9 + bi); //one point on the line @ x = 9
return y; //the local one, bi is global
}
```

This function computes smoothed-out values for the four control values computed in `waypoint_drive()`. It only uses one column of values in stack4 as specified by the formal parameter column. The method that it employs is to first make up a sequential list of values for an x-axis comprising 10 integers. It then sums those integers. This is not a constant since the user can choose to select more or less than 10 previous values. It then accesses the first 10 values in stack4 and sums those values. It then computes the product of the x-values and corresponding stack4 values. It then computes the square of each of the `stack4` values. With this information, it computes the slope of the best fit RMS line passing through the `stack4[][]` values. It similarly computes the y-intercept of that line. It returns the slope in `m` and the intercept in `bi`, both of which are globals intended for such purposes. In this way, this function averages out the jitter that may be present in the control and command values and preserves any trend that may be forming over the last 10 `locate()` cycles. Using the y-intercept as the representative value for the distance anticipates any upward or downward trend. This function also returns a y-value that is historically 10 cycles in the past with respect to the smoothed-out value reported in `bi`. This function is no longer used by the program since jitter and outliers in the control and command values stream are now too rare to be considered separately. In previous versions of the code that smoothed out the `stack4[][]` values, it was useful.

```
float find_diff(int column) { //best fit RMS line to points in
stack4[10][c]
int i, n = 10; //our durable indexes
float xlist[10] = {0,1,2,3,4,5,6,7,8,9};//the stack has sequential
values
float sx = 0, sy = 0, sxy = 0, sx2 = 0, y; //needed for RMS
calculation
for(i = 0;i < n;i++) { //fill sums with n values
sx = sx + xlist[i]; //a constant
sy = sy + stack4[i][column];//column = the mike number: 0, 1, 2, ...
sxy = sxy + xlist[i]*stack4[i][column];//the big player
sx2 = sx2 + xlist[i]*xlist[i]; //sum of x-squares, a constant
}
m = (float)((sxy - sx*sy/n)/(sx2-sx*sx/n)); // formula for slope
bi = (float)(sy/n - m*sx/n); // formula for intercept
y = (float)(m*9 + bi); //one point on the line @ x = 9
return y; //the local one, bi is global
}
```

This function computes smoothed-out values for the heading values found on the `stack` reserved for `xe`, `ye`, and `ze` values. It only uses one column of values in the stack (the third, number 2), so there are no formal parameters. Since `ze` is presented as a principal value, it could jitter between plus or minus p, and thus be difficult to "average" in a meaningful way. The function does the nearly same process twice for a pair of `ze` components: a `sin(ze)` and a `cos(ze)`. The method that it employs is to first make up a sequential list of values for an x-axis comprising 10 integers. It then sums those integers. This is not a constant since the user can choose to select more or less than 10 previous values. It then accesses the first 10 values in the stack and sums the sine or cosine of those values. It then computes the product of the x-values and corresponding sine or cosine stack values. It then computes the square of each of the x-coordinate values. With this information, it computes the slope and intercept of the best fit RMS line passing through the transformed `stack[i][2]` values. It ignores the slope in m, but saves the intercepts in bsin and bcos, both of which are local variables. It finally computes and returns an averaged `ze` from `atan2(bsin, bcos)`. In this way, this function averages out the jitter that may be present in the differences in xe and ye values and preserves any trend that may be forming over the last 10 `locate()` cycles. Using the y-intercepts as the representative value for the distances anticipates any upward or downward trend. This function is no longer used by the program since jitter and outliers in the `xe` and `ye` values stream are now too rare to be considered separately. In previous versions of the code that smoothed out the `stack[i][2]` values, it was useful.

```
float find_ze() { //best fit RMS line to points in stack[10][2]
int i, n = 10; //our durable indexes
float xlist[10] = {0,1,2,3,4,5,6,7,8,9};//the stack has sequential
values
float sx = 0, sy = 0, sxy = 0, sx2 = 0; //needed for RMS calculation
double bsin, bcos, bz; //new variables
for(i = 0;i < n;i++) { //fill sums with n values
sx = sx + xlist[i]; //a constant
sy = sy + sin(stack[i][2]); //parameterize ze with sin
sxy = sxy + xlist[i]*sin(stack[i][2]); //the big player
sx2 = sx2 + xlist[i]*xlist[i]; //sum of x-squares, a constant
}
m = (float)((sxy - sx*sy/n)/(sx2-sx*sx/n)); // formula for slope
bsin = (float)(sy/n - m*sx/n); // formula for intercept
sx = 0, sy = 0, sxy = 0, sx2 = 0; //needed for RMS calculation
for(i = 0;i < n;i++) { //fill sums with n values
sx = sx + xlist[i]; //a constant
sy = sy + cos(stack[i][2]); //parameterize ze with cos
sxy = sxy + xlist[i]*cos(stack[i][2]); //the big player
sx2 = sx2 + xlist[i]*xlist[i]; //sum of x-squares, a constant
}
m = (float)((sxy - sx*sy/n)/(sx2-sx*sx/n)); // formula for slope
bcos = (float)(sy/n - m*sx/n); // formula for intercept
bz = atan2(bsin,bcos); //restore the fitted ze
return (float)bz; //the intercept is our heading
}
```

This function sets up a solution for the heading based upon `xe` and `ye` values loaded into the `stack[][]` at columns 0 and 1, respectively. Two points in the plane are needed. These are calculated by passing a line through the most recent 10 values of the `xe` and `ye`. The line for heading calculation is based on the averaged `xe` and `ye` values (their respective `bi`-values) and the pair appearing 10 positions back in the stack. The long distance back in time gives the calculated heading a longer vector with far less jitter than would otherwise be found. This routine is no longer used in the program.

```
float get_ze(){ //pass a line through each xe & ye stacks
float x1, y1, x2, y2; //obtain 2 points
x2 = find_dir(0); //get end of the xe line
x1 = (float)bi; //get start of the xe line
```

```
y2 = find_dir(1); //get end of the ye line
y1 = (float)bi; //get start of the ye line
ze = get_heading(x2, y2, x1, y1); //finally, get ze from that
//if(principal(tar-ze) > PI/2) ze = principal(ze + (float)PI);//check
for flips
//if(principal(tar-ze) < -PI/2) ze = principal(ze-(float)PI);//big
and small
return 0;
}
```

This function is used to compute a good heading value from differences in the triangulated `xe` and `ye` values found on the `stack[][]`. First, the routine computes the differences between values in each of the 0 and 1 columns of the `stack[][]` as far as 10 deep in past history. From these, it computes a pair of average values representing the `xe` and `ye` over that time interval. Any values on the stack that are far different from the average are considered "outliers" and need to be removed. This removal occurs next by re-checking each difference between a given stack value and its next earlier one (given by `i` and `i+1`, respectively.) A value that is greater in magnitude than twice the average is replaced by the average itself. These are all summed into `ex` and `ey`, respectively. The heading is then computed based on the smoothed historic values only. This routine is called by `locate()` and is successful in finding good headings from otherwise jittery input data. Since it reaches into history 10 events deep, `ze` is not available until 10 cycles have passed. For this reason, new waypoint line segments begin with a short `startout()` move, just to fill the `stack[][]`.

```
float smoother(){ //get ze from ten smoothed historic data points
int i;
float dx = 0, dy = 0, ex = 0, ey = 0, delt;
for(i = 0;i < 10;i++) { //reach back ten events
dx = dx + stack[i][0]-stack[i + 1][0]; //get delta x
dy = dy + stack[i][1]-stack[i + 1][1];} //get delta y
dx = dx/10; dy = dy/10; //get averages
for(i = 0;i < 10;i++) { //re-do the scan
delt = stack[i][0]-stack[i + 1][0]; //get delta x
if(2*abs(delt) > abs(dx)) ex = ex + dx; //outlier! use average
else ex = ex + delt; //use actual value
delt = stack[i][1]-stack[i + 1][1]; //get delta y
if(2*abs(delt) > abs(dy)) ey = ey + dy; //outlier! use average
else ey = ey + delt;} //use actual value
ex = ex/10; ey = ey/10; //get averages
ze = atan2(ey,ex); //magic heading
return 0;
}
```

This function finds the current values of `xe`, `ye`, and `ze` based on the received distances from two receivers and their respective locations in the plane space of the robot. There are two formal parameters: an estimate of the total number of cycles or events needed to complete the current waypoint line segment, `j`, and the number of events computed to correspond to the point where the waypoint line segment crosses a line between the two receivers, `jstar`. These two values will determine whether we are approaching the intersection or having just passed it. This event determines which of the calculated navigational triangles to use in finding `xe` and `ye`. First, the function checks the values of `xe` and `ye` for zeros. Zero values here would indicate that a `waypoint_drive()` is being initiated, and an estimate for these values is needed to determine which triangle to use. The routine stops the robot if it were moving for any reason. Then it requests estimated values for `xe` and `ye` from the user to get started. The routine also check to see if the identities of the receivers have been set. If the two subscripts representing them are equal, they are most likely zero or not correct. If so, the routine asks for these values so that the correct receivers are used for the patch to be mowed. The routine pauses so that there will be valid data coming from these receivers.

This function may begin with known values of `xe`, `ye`, `ca`, and `cb`. If the robot is not moving at this time, then the most accurate starting point is computed by calling `get_mode()`. A `triangulate()` is then done to find new values for `xe` and `ye`. If the triangulation is successful, the variable ret is 0, otherwise, it will be a 1. Three cases are examined next: whether there is a crossing point between two receivers, whether the robot is heading towards this intersection, or whether it has passed it. For comparison with the calculated pairs of `xe` and `ye` values, dummy variables xc and yc are filled. If no crossing exists, then jstar will be zero, and the comparison values can be simply `xe` and `ye`. Otherwise, there will be a crossing and xc and yc are loaded with the starting value of the waypoint line segment or the destination value. With all possible combinations taken into account, the pair of values `xe` and `ye` that are closest to the comparison values, xc and `yc`, are selected.

If the `triangulate()` function should fail or if the robot is within a few inches of the line connecting two receivers, then the next `xe` and `ye` values are extrapolated. The user gets a beep if this happens. Finally, the new `xe` and `ye` values are put onto the `stack[][]`, ze is calculated and put on the `stack[][]` and the current control values are stored on `stack4[][]`. In this way, the user can retrace recent history for all pertinent variables should the need for analysis of the robot moves become necessary. In this case, the user would examine the file named "`work`," which is automatically written with each new waypoint line segment.

```
int locate(int j, int jstar) { //code for x-y-z location from system
dme data
float xc, yc; //comparison location values
if(xe ==0){ //if no estimate of location, ask for it
stop_servo(); //stop for a moment
printw(" What is xe? "); Beep(2000,20);//screech
getstr(value); //manual guess
if(!strcmp(value,que)) return 1;//quit now?
xe = (float)atof(value); //but use it
servo_go(); //resume
}
if(ye ==0){ //if no estimate of location, ask for it
stop_servo(); //stop for a moment
printw(" What is ye? "); Beep(2000,20);//screech
getstr(value); //manual guess
if(!strcmp(value,que)) return 1;//quit now?
ye = (float)atof(value); //but use it
servo_go(); //resume
}
if(ca==cb){ //no good if both are the same
printw("Using..."); //so ask for new ones
getstr(value); //manual input
if(!strcmp(value,que)) return 1;//quit now?
ca = (int)atof(value); //but use it
getstr(value); //manual input
if(!strcmp(value,que)) return 1;//quit now?
cb = (int)atof(value); //no, use it
Sleep(80); } //let call-back catch up to have da & db
if((int)mov[0]==ZERO && (int)mov[1]==ZERO) get_mode();//do job while
stopped
ret = triangulate(X[ca],Y[ca],da,X[cb],Y[cb],db);//use the current
receivers
```

```
if(jstar==0) {xc = xe; yc = ye;} //normal case, jstar =0, no
intersection
else if(j < jstar) {xc = xs; yc = ys;} //starting case, j < jstar
else {xc = xd; yc = yd;} //finishing case, j > jstar
if(((xc-xup)*(xc-xup) + (yc-yup)*(yc-yup))<
((xc-xum)*(xc-xum) + (yc-yum)*(yc-yum))
&& xup > 0 && yup > 0) { //choose the + location nearest xc, yc
xe = xup; ye = yup;} //we can come back, too
else {xe = xum; ye = yum;}
if(abs(dist2line(X[ca],Y[ca],X[cb],Y[cb],xe,ye)) < 6 || ret==1) {
extrapolate(); Beep(200,10);} //an error makes us rely on the plan
//stop_servo(); //stop on error!
//getstr(value);}
coord[0] = xe; pushdown(0); //raw xe value based on da, db
coord[1] = ye; pushdown(1); //raw ye value based on da, db
smoother();
coord[2] = ze; pushdown(2); //smoother ze value based on xe, ye
push4(); //keep run values in sync
return 0;
}
```

This function returns the arctangent of a line given by a pair of points on the line. It has formal parameters that pass the point set to it, and it makes a single correction so that the value returned is a principal value. This value is not checked.

```
float get_heading(float x1, float y1, float x2, float y2) {//not for
ze!
double heading; //useful to know the angle of any line
heading = (double)atan2((double)(y2-y1),(double)(x2-x1));
if(heading > PI) heading = heading-2*PI; //correct for step change
if(heading < -PI) heading = heading + 2*PI;
return (float)heading;
}
```

Central to Field Robot Navigation is this `triangulate()` function. Its purpose is to compute estimated x and y locations, `xe` and `ye`, from the distances reported by the `InputHandlerE`. For generality, its formal parameters are as follows: the known x and y locations of a first receiver, the distance reported by this receiver, the known x and y locations of a second receiver, and the distance reported by this receiver. The distances from these sources are used to set variables a and b, respectively. The distance between the receivers is calculated for variable c, and it should match a known value independently. This value gives a way of checking the locations of both receivers. If the reported distances are incorrect, the routine prints the three values and returns with an error code = 1. Next, a geometric value, s, is calculated from standard triangle-solving formulae. This number is the "average" of the three sides. From this value, the half-sines of the angles opposite sides a and b are calculated. To obtain the most accurate values possible, the range of the half-sines are checked. We are free to use either the "a" half-sine or the "b" half-sine. The one that would give the most accurate arcsine is chosen by the test. Since the triangle has a general orientation in the plane, its "tilt," `th`, is then calculated from the known positions of the receivers. An error could occur in the calculation if an impossible triangle is presented to the code. The error would appear as a half-sine greater than unity. If this happens, the routine reports the three offending numbers and returns with an error code = 1. A standard geometric formula is then used to calculate a pair of alternative x and y locations, as shown in Figure 6.36. This calculation occurs only once with the better of the two arcsine values. The reason for the calculation of a pair of x and y values lies in the inherent ambiguity of the result: there are two possible triangles computed and we do not know which one to use. We have labeled these as x-unknown-plus and y-unknown-plus, or x-unknown-minus and y-unknown-minus. All four values are stored as globals for the `locate()` routine to evaluate and use.

```
int triangulate(float x1,float y1,float d1,float x2,float y2,float
d2) {//find x-y
double a, b, c, SHA, HA, A, SHB, HB, B, s, th; //from mike locations
bool useb = true; //alt method based on the
//solve navigational triangle given distances d1 and d2//accuracy of
the asin()
a = d1; b = d2; c = sqrt((double)((x1-x2)*(x1-x2) + (y1-y2)*(y1-
y2)));//set up lengths
if((a + b) < c) { //check for a viable triangle
printw(" Cannot Triangulate! "); //ooops, trouble in geometry land
printw(" a = %g, b = %g, c = %g. ",a,b,c); //report the news
return 1;} //unrecoverable error
s = (a + b + c)/2; //check for accuracy of arcsin next:
SHA = (sqrt((s-b)*(s-c)/(b*c))); if(SHA > .7) useb = true;
SHB = (sqrt((s-a)*(s-c)/(a*c))); if(SHB > .7) useb = false;
th = atan2((double)(y1-y2),(double)(x1-x2)); //tilt of solved
triangle
//printw(" th = %5.2 f. ",th);
if(SHA > 1 || SHB > 1) {printw(" Bad Arcsin!! ");
printw(" a = %g, b = %g, c = %g. ",a,b,c); return 1;}
if(useb) { //use B for basic analytic geometry
HB = asin(SHB); B = HB*2; //printw(" B = %5.2f;",B);
xup = (float)(x1 + a*cos(PI-(B-th)));
yup = (float)(y1 + a*sin(PI-(B-th)));
xum = (float)(x1 + a*cos(PI-(B + th)));
yum = (float)(y1-a*sin(PI-(B + th)));}
else { //else, use A for basic analytic geometry
HA = asin(SHA); A = HA*2; //printw(" A = %5.2f;",A);
xup = (float)(x2 + b*cos(A + th));
yup = (float)(y2 + b*sin(A + th));
xum = (float)(x2 + b*cos(A-th));
yum = (float)(y2-b*sin(A-th));}
//printw(" a = %5.2f, b = %5.2f, c = %5.2f, xup = %5.2f,
yup = %5.2 f. ",a,b,c,xup,yup);
return 0; //four values returned in globals
}
```

This function reads a text file stored in the folder designated with the name of the program, namely `PlanD`, and at the same level as the Visual Studio solution file. It is set to be no longer than 100 lines, but this is easily changed to any reasonable value. The function requests the file name from the user screen and opens a `workfile` for reading. An error in opening the file results in an error message to the user and termination of the routine. The function will read values into a special list called `point_list[][]`. This list comprises each of the waypoints that the robot will traverse in the *Mow Lawn* basic function. The order of the columns is important and fixed by the function as: the starting point of a waypoint segment, the destination point of this segment, the velocity to be used during this move, the identity of the first receiver, and the identity of the second receiver. The function clears the entire list before reading new values since the termination of the list is sensed by `waypoint_drive()` as a zero where a waypoint would normally be found. The `fscanf` command isused to read the values, and it stops when the first leading zero is found. The function then prints the list to the screen for user verification. This function is called only by `waypoint_drive()` in preparation for mowing a patch of lawn.

```
int fileread(){ //read waypoint values before we move about
char fname[20];
int i, j = 100; //our indexes
printw("Enter filename:"); getstr(fname); //identify file
if(fname! = NULL) { //standard file command
workfile = fopen(fname,"r"); //read
if(workfile==NULL) {printw("Some kind of workfile error.");
return 0;}
}
if(fname! = NULL) { //standard file command
for(i = 0;i < j;i++) { //clear the waypoint list
point_list[i][0] = 0;
point_list[i][1] = 0;
point_list[i][2] = 0;
point_list[i][3] = 0;
point_list[i][4] = 0;
}
for(i = 0;i < j;i++) {
fscanf(workfile,"%d\t%d\t%d\t%d\t%d\n", //get a new list
&point_list[i][0],&point_list[i][1],&point_list[i][2],
&point_list[i][3],&point_list[i][4]);
if(point_list[i][0]==0) break;}
for(i = 0;i < j;i++) { //to be sure,
printw("%d\t%d\t%d\t%d\t%d\n", //show the list
point_list[i][0],point_list[i][1],point_list[i][2],
point_list[i][3],point_list[i][4]);
if(point_list[i][0]==0) break;}
}
return 0;
}
```

## 7.3 Operation: The Calls that Make the Robot Move and Stop

These functions are used primarily for control of the robot on a real-time basis, notwithstanding the delays inherent in USB communications and Windows™ in general. These functions begin with the system calling `main()`. From there the user can select several functions. For normal *Mow Lawn* operation, the user first calls `locate()` to determine the x, y position of the robot. (Remember, the heading, z, will not be known until the robot has moved a short distance.) If the system is just starting up from "control F5," `locate()` will request an initial guess and the ID numbers of the receiver locations being used for this run.

Finally, the user invokes `waypoint_drive()` with a single "w" keystroke. This function directs all robot movements. The user is asked for the name of the text file describing the locations used for the lawn patch about to be mowed. Then `waypoint_drive()` takes over and runs the robot until all the waypoint points have been successively followed. At any time, the user can touch the "q" key, and the program will quit, stopping the robot. A re-start requires simply another invocation of `waypoint_drive()`. Invoking `locate()` is optional at this point.

The function to follow this introduction sets the servo command values to `ZERO` so that the robot motors stop. The `ZERO` value itself needs to be found through experiment with the pulse width modulation (PWM) driver. All four servo channels are set to zero, even channel 4 which carries the sync information. This function is called by many others: `move_servo()`, `locate()`, `startout()`, `main()`, and `waypoint_drive()`.

```
int stop_servo(){ //for any time the robot needs to halt
mov[0] = ZERO; mov[1] = ZERO; mov[2] = ZERO; mov[3] = ZERO; //set
values
CPhidgetAdvancedServo_setPosition(SRV, 3, mov[3]);//used for sync,
don't care
CPhidgetAdvancedServo_setPosition(SRV, 2, mov[2]); //for data display
CPhidgetAdvancedServo_setPosition(SRV, 1, mov[1]); //board-side motor
CPhidgetAdvancedServo_setPosition(SRV, 0, mov[0]); //switch-side
motor
screen_init(); //get the screen ready
update_display_servo(mov); //screen1
printw("!"); //show on the screen that we halted
return 0;
}
```

This next function accepts the servo command values in the variables `mov[0]`, `mov[1]`, and `mov[2]`, so that the robot motors and display can be enabled by a subsequent call to the Phidgets library. The values themselves needs to be found through the calling code prior to `servo_go()`. This function is called by two others: `main()` and `waypoint_drive()`.

```
int servo_go() { //universal signal to GO
CPhidgetAdvancedServo_setPosition(SRV, 0, mov[0]); //do it now
CPhidgetAdvancedServo_setPosition(SRV, 1, mov[1]); //do it now
update_display_servo(mov);
return 0;
}
```

This next function accepts a single servo command value from the keyboard at a prompt "`number`?" It sets this value into the servo channel preselected by the code itself. Since it is used for experiment, there is no problem in setting the channel in the code itself. We illustrate with the third channel (2). The command is enabled by a subsequent call to the Phidgets library. This function is called only by `main()`, and it has no terminal condition except by exiting the program.

```
int test_servo() { //universal signal to GO
double j;
while(1){ //travel dist tolerance
if(getch()==113) break; //e-stop
printw(" number? "); Beep(2000,20); //screech
getstr(value); //manual guess
if(!strcmp(value,que)) return 1; //quit now?
j = atof(value); //no, so use it
CPhidgetAdvancedServo_setPosition(SRV, 2, j);
}
return 0;
}
```

This next function accepts two servo command values from the keyboard at a prompt "`Servo (#) =`." It sets these values into servo channels 0 and 1 as preselected by the code itself. Since it is used for experiment, there is no problem in restricting setting the channel to the two drive motors. The command is enabled by subsequent calls to the Phidgets library. This function is called only by `main()`, and it has no terminal condition except by typing a "q-enter."

```
int move_servo(){ //low-level TEST function for set-up
int i;
stop_servo(); //reset to zero, first
while(1){ //do until we shut it down
for(i = 0; i < 2; i++){ //we'll use 2 servos for now
```

```
printw("Servo %d = ",i); //prompt for input
getstr(value); //get the input value
mov[i] = atof(value);} //sends a double
CPhidgetAdvancedServo_setPosition(SRV, 0, mov[0]); //do it now
CPhidgetAdvancedServo_setPosition(SRV, 1, mov[1]); //do it now
screen_init(); //do these when you have something to show
update_display_servo(mov);
if(!strcmp(value,que)) break; //try this out, needs 'q' and CR
} //for typed value on input
stop_servo(); //stop on quit
return 0;
}
```

This next function computes the number of `locate()` cycles during which there may be no distance measurements possible by the `InputHandlerE` interrupt routine. The condition may occur if the robot is located temporarily in the acoustic shadow of a tree or bush. It also may occur when the robot approaches the line connecting two receivers, during which `triangulate()` is not possible. The routine begins by retrieving the commanded velocity of the robot vel for the current waypoint line segment as found by the `point_count` value. The origin of the commanded velocity is the file composed by the user and called at the start of `waypoint_drive()`. The routine then estimates the number of cycles that the robot should take to complete the entire line segment at that velocity. The line length is computed from the start and destination points defining the line segment and stored in `L`. The estimated number of total cycles is then computed, stored in `je`, and displayed for the user. This estimate is used only to check on the validity of the estimate. The location of the intersection of two lines is then computed. First, the slope and intercept of the waypoint line segment is found and stored in `m1` and `b1`, respectively. Secondly, the slope and intercept of the line segment connecting the two receivers in use is found and stored in `m2` and `b2`, respectively. These determine an intersection point stored in `xi` and `yi`. The routine then computes the distance from the start point to the intersection point (in inches) and converts it to the estimated number of cycles, `jstar`, needed to reach that point. This value is used by `waypoint_drive()` to direct the robot during the `triangulate()` "blackouts." It is also used to determine when to select the alternative triangle computed by `triangulate()` during passage through this area.

```
int get_jstar(){ //find the number of steps until we flip the
triangle
int jstar = 0, je = 0; //estimate j
double L, m1, m2, b1, b2, midlength; //line parameters
vel = point_list[point_count][2]; //slowly, for now
if(vel==0) return 0; //check for end!
L = sqrt((xs-xd)*(xs-xd) + (ys-yd)*(ys-yd)); //waypoint line length
je = (int)(30.*L/abs(vel)); //magic number HERE
printw("je = %d. ",je); //compare with actuals
if(get_line(xs, ys, xd, yd) > 0) return 0; //find it
m1 = m; b1 = bi; //got it
if(get_line(X[ca], Y[ca], X[cb], Y[cb]) > 0) return 0;//get receiver
line
m2 = m; b2 = bi; //got it
if(get_intersection(m1, b1, m2, b2) > 0) return 0;//where they
cross = jstar
midlength = sqrt((xi-xs)*(xi-xs) + (yi-ys)*(yi-ys));//inches to jstar
if(midlength < L) jstar = (int)(30.*midlength/abs(vel)); //magic
number HERE
else jstar = 0; //no intersection
return jstar; //if it exists
}
```

This next function estimates values for xe and ye (from which ze will be derived) when the triangulate() function fails or there are no good distances returned from InputHandlerE. The condition may occur if the robot is located temporarily in the acoustic shadow of a tree or bush. It also may occur when the robot approaches the line connecting two receivers, during which triangulate() is not possible. The routine begins by computing the component of the robot's next incremental motion along the target line between two waypoints. Since the velocity is given as a percentage of the top speed, it uses a scale factor of 0.1 to bring the net value into agreement with the expected movement. Values for both the x and y directions are computed, and then added to the current values stored on the stack[][] in the xe and ye columns, respectively. A message to the user indicates when this step has been taken. It will be accompanied by a Beep() if there is a missed distance reading from InputHandlerE.

```
int extrapolate() { //use ten historic points to find the nexts in
line
float delx = 0, dely = 0; //for the ideal delta for x & y
delx = (float)(cos(tar)*abs(vel/10.)); //estimate from tar
dely = (float)(sin(tar)*abs(vel/10.)); //estimate from tar
xe = delx + stack[0][0]; //add it to the last x-value
ye = dely + stack[0][1]; //add it to the last y-value
printw("Extrap'd xe = %d, ye = %d. ",(int)xe,(int)ye);
printw("delx = %5.2f, dely = %5.2 f. ",delx,dely);
return 0;
}
```

This next function moves the robot about 10–50 cm in the direction given by the current velocity value, vel. This move is useful for filling the stack with location values while not commanding any turns. The velocity is given as a percentage of the robot's top speed, so it is scaled. Also, the value passed to the servo control is offset by the center or ZERO control position. The robot moves a varying distance depending on the speed setting, but the number of events is set to 12. The code is set to Sleep(80) between events, so that it matches the interrupt routine, InputHandlerE. A brief stop of the robot is coded in case the calling sequence does not have such a safety feature. This function is called by waypoint_drive().

```
float startout() { //wherever you are, move to fill the stack
int j;
vel = point_list[point_count][2]; //reset velocity to be sure
mov[0] = vel/2 + ZERO; //set up servo command, switch-side
mov[1] = vel/2 + ZERO; //set up servo command, board-side
servo_go(); j = 0; //move out! if needed
for(j = 0;j < 12;j++) {
Sleep(100); locate(0,0);} //get 12 initial values into the stack
stop_servo(); //don't go 'way now
return 0; //the last heading found = ze
}
```

This next function checks the d2line stack to determine if the errors are so large that the robot cannot be following the waypoint line segment. If this is the case, we consider the robot to be lost and immediately stop its movement, and report the condition to the user screen. The function checks the current five values of the stack4[i][3] and if they are all greater than 24 in., then the lost() condition is satisfied. The code stops the robot and does a locate(), which will in turn, call get_mode() for an accurate reading. The screen displays the condition and beeps. The displays are updated and control returns with an error code = 1. This function is no longer used by the program, but was useful during servo development.

```
int lost(){
//return 0; //subtract it for testing
if( abs(stack4[0][3]) > 24 &&//look to the d2L column
abs(stack4[1][3]) > 24 &&//check on growth of errors
abs(stack4[2][3]) > 24 &&//
abs(stack4[3][3]) > 24 &&//
abs(stack4[4][3]) > 24 ){ //if 5 big d2L errors, then stop
stop_servo(); //quickly
locate(0,0); //do a mode find and quit
printw("\n LOST!!"); Beep(500,50);
update_display3();
update_display4();
return 1;} //locate will filewrite this
return 0;
}
```

This next important function drives the robot according to the waypoints saved into a text file. Several lines have been commented-out to preserve some of the history of this pivotal function for reader analysis. There are several "magic" values coded into the function that are needed for servo control of the robot we built according to the functional and schematic figures given in the previous chapters of this book. There are just two main loops: an outer loop that indexes with each new waypoint line segment, and an inner loop that controls events as the robot moves along that segment. As each segment is completed, filewrite() is called to memorialize many of the variables according to a time history as stored on various stacks. The "q" key can be hit any time to halt the robot and advance to the next waypoint or simply quit the function. Some subfunctions are done in-line that may well have been taken out of the program flow, and these will be noted carefully as the description follows the code.

A number of declarations are made for local variables. The `(float)` values are needed to control the servo motors on the robot as it encounters disturbances in an uneven field. These have been calculated from control theory and validated by experimental design. We recognize that the derivative of `d2line` needs to be calculated for best anticipation of accelerations by the system, but differences were used instead and found to be satisfactory. The average error values for each stack of 60 values were used for objective functions as these numbers were developed for best quantitative control. The stacks of "`xyz.txt()`" values were very useful in determining the best functions for removing jitter from the distance readings and the leverage effect they would have in computing the heading. There are no "`goto`" statements in the code, and no intricate or unusual code meanings.

After the initializing steps are taken, the function reads in the entire list of waypoints to be followed. A counter to the current waypoint line segment is set to "1," and the identity of the receivers extracted from the list. The code waits for the interrupt routines to obtain new values, and the initial location of the robot is attempted. Help from the user may be needed at this point. Before entering the first waypoint loop, "quit" is tested. An index, n, is kept for emergency termination of the loop for any reason. The start and destination points of the waypoint line segment are loaded. The loop terminates with a break if the current waypoint value is zero. The user is prompted that the next line segment is available, and the heading of this line is computed and displayed. The possible presence of a crossover point in this line segment is checked. Then the finishing condition for this line segment is computed. This is done by finding a line perpendicular to the waypoint line segment that runs through the destination point. The distance to this line is computed and the sign saved. The robot is stopped (if it were moving), and the "q" checked again. The velocity is extracted from the list. It is given as a percentage of the top speed of the robot. A limit on this speed is set so that turns will not be so tight as to damage the turf. If the stack of x and y values is empty, or the robot changes direction by 180°, then the robot makes an initial move for a short distance in a straight line. The servos are set for a soft start by calling the `phidget21.h` library and the inner loop begins.

The function of the inner loop is to control the path of the robot along the waypoint line segment extracted by the outer loop. The terminating condition normally is that the distance to go, `d2go`, for the terminating perpendicular line is less than a few inches. Alternatively, any anomaly detected could set `n` unequal to zero and the loop would break. A number of keypress checks are made to stop or continue the robot motion. Then `d2go` is computed and printed along with the current number of events or `locate()` cycles completed. The lateral `error` of the robot with respect to the waypoint line segment is computed and its sign saved. The `error` in heading is computed and saved. The derivative of the error is calculated. Then the control law is implemented, deriving the value, `delta`, to be applied to steer the robot into alignment as needed. Limits to the error are applied and the status of `triangulate()` is checked. If there is no valid findings of `xe` and `ye`, then the `delta`-value is set to zero. This means there can be no turns during a "blackout," and hence no integration of errors. An error measure is computed for history, and the moves for the drive wheels are set. These are checked against the PWM board limits. The display is updated, and the robot made to travel. The code then sleeps for 80 ms while the interrupts take place. The event counter is then incremented and, the `locate()` function invoked. This ends the inner loop.

The outer loop then changes the parameters of the drives for a quick stop, if needed, and the robot brought to a brief halt. The average error is computed for history, the waypoint line segment counter advanced to the next waypoint, and the recent history dumped to a file for optional analysis. This ends the outer loop. If an error occurred, the history file is dumped. The robot is halted and the system beeps. The parts of the code that have been commented-out show how various changes were considered in smoothing out the action of the robot, or diagnoses of problems were aided. No code is perfect, so we expect changes in the future as we cut more grass.

```
int waypoint_drive() {
int i, j, jstar, k = 0, sign = 0, dsign = 0,
esign = 0,limit = 1;//indicies & helpers
float kp = (float)0.16, kd = (float)25.6, ka = 40, olderror = 0,
deriv = 0;
float oldderror = 0, ddiff = 0, derror = 0, d2go;
float xf, yf; //for arcs & lines
//xe = 1200; ye = 300; //CHEAT!! Outside starter value
//xe = 220; ye = 210; //For inside tests, give a starter value
//xe = 130; ye = 10; //For doorway tests, give a starter value
fileread(); //get waypoints
point_count = 1; //first waypoint is here
ca = point_list[point_count][3]; //fetch ca, the 1st receiver number
cb = point_list[point_count][4]; //fetch cb, the 2nd receiver number
Sleep(100); locate(0,0); //get mode location, accurately
stop_servo(); getstr(value); //for now, just in case
if(!strcmp(value,que)) return 0; //try this out, needs 'q' CR
while(n==0) { //for all waypoints
xs = (float)point_list[point_count-1][0]; //start point for first
test line
ys = (float)point_list[point_count-1][1];
xd = (float)point_list[point_count][0]; //where shall we go next?
yd = (float)point_list[point_count][1]; //in 2 dimensions
if(xd==0) break; printw("\n Next waypoint at %d, %d.
```

```
",(int)xd,(int)yd);
printw(" point_count = %d. ",point_count);
tar = get_heading(xs,ys,xd,yd); //corrected for step change
printw(" target = %6.2 f. ",tar);
jstar = get_jstar(); //find crossover distance
printw(" jstar = %d. ",jstar);
get_perp(xs,ys,xd,yd); //find finish-line through xd,yd
xf = xd + 1; yf = (float)(xf*mp + bp); //any x will do, solve for y
if(dist2line(xf,yf,xd,yd,xs,ys) > 0) sign = 1;//set side sign
else sign = -1; //get side of finish line set
stop_servo(); getstr(value); //read-up, me hearties!
if(!strcmp(value,que)) return 0; //try this out, needs 'q' and CR
clear(); //each waypoint gets a fresh screen
vel = point_list[point_count][2]; //slowly, for now
limit = (int)abs(vel/3); //accomodate speeds
if(stack[10][0]==0) startout(); //fill up stack initially
if(point_list[point_count-1][2] = -point_list[point_count][2])
startout();
//if we just commanded a reversal, add a little heading-finder
d2go = 1000; j = 0; arg1 = 0; //reset indices
CPhidgetAdvancedServo_setAcceleration (SRV, 0, 1000);//soft start
CPhidgetAdvancedServo_setAcceleration (SRV, 1, 1000);//soft start
while(d2go > 5 && n==0){ //travel dist tolerance
if(getch()==113) {stop_servo(); n = 1; getstr(value);} //e-stop
if(getch()==103) {servo_go();} //e-stop
if(getch()==115) {stop_servo();} //e-stop
d2go = (float)dist2line(xf,yf,xd,yd,xe,ye)*sign;//from a normal
finish line
printw(" %d. ",(int)d2go); //keep tabs on our progress
printw("j = %d. ",j); //keep tabs on our progress
//oldderror = d2line; //get a difference
d2line = (float)dist2line(xs,ys,xd,yd,xe,ye);//get lateral error
if(d2line > 0)dsign = 1;
else dsign = -1;
//d2line = (float)((d2line + stack4[0][3] + stack4[1][3])/3.);//movin
g average
//ddiff = (float)(d2line-oldderror); //compute difference/deriv
//ddiff = (float)((ddiff + stack[0][3] + stack[1][3])/3.);//moving
average
//coord[3] = ddiff; pushdown(3); //save it for history
olderror = error; //for just a difference
error = principal(tar-ze); //get new angular error
if(error > 0)esign = 1;
else esign = -1;
//error = (float)((error + stack4[0][2] + stack4[1][2])/3.);//moving
average
diff = (float)(error-olderror); //compute difference/deriv
//diff = (float)((diff + stack4[0][0] + stack4[1][0])/3.);//moving
average
delta = (float)(d2line*kp + error*kd + diff*ka); //compute control
law
if(delta > limit)delta = (float)limit; //cap it in case of jumps
if(delta < -limit)delta = -(float)limit;
if(ret==1) delta = 0; //trouble, so don't turn!
arg1 = arg1 + (float)abs(d2line); //collect an error measure
mov[0] = vel/2 + ZERO + delta*manual; //set up servo command, switch-
side
mov[1] = vel/2 + ZERO-delta*manual; //set up servo command, board-
side
for(i = 0;i < 2;i++){ //keep boundaries enforced
if(mov[i] < 40) mov[i] = 40; //don't ask for less than possible
if(mov[i] > 160) mov[i] = 160;} //or more than possible
screen_init(); update_display4(); //show all run-time data
servo_go(); //start your engines, already!
Sleep(80); //wait for next input cycle
```

```
j++; //save distance & position, too
locate(j,jstar); //n = lost(); //see where we got to, lost?
//if(j%10==0){ //look for clews!
// stop_servo();filewrite();
// getstr(value);}
//cd--; if(cd < 0) cd = 0; //countdown for backup
} //end of d2go while loop
CPhidgetAdvancedServo_setAcceleration (SRV, 0, 10000);//quick stop
CPhidgetAdvancedServo_setAcceleration (SRV, 1, 10000);//quick stop
stop_servo(); //for a millisecond
printw(" j = %d ",j); arg1 = arg1/j; //collect avg error for printing
point_count++; //bump waypoint pointer
filewrite(); //store data gathered during run
//xyzwrite(); //store xyz data gathered during run
} //end of while loop
if(n==1) filewrite(); //store data gathered if lost
stop_servo(); Beep(500,500); //Success!!
return 0;
}
```

This auxiliary function gathers the time of day in milliseconds for the user when diagnosing problems in the system. This function is only called by `main()`.

```
int time_reads(){
//int i;
GetSystemTime(&time);
millis = time.wMilliseconds;
printw(" %u ", millis); //what time is it??
locate(0,0);
GetSystemTime(&time);
millis = time.wMilliseconds;
printw(" %u ", millis); //what time is it??
return 0;
}
```

This independent function is initiated by the system in Visual Studio by touching the "`control F5`" keys together. It stands as the starter for the USB connection functions and the closer for those functions. It initiates a number of functions automatically to load global variables with values. Lines that are commented-out may be selected for special conditions, such as indoors or outdoors, depending on circumstances. In between, it responds to a single keystroke for various commands, as detailed in each case, below. The basic mechanism is a simple `switch-case` construct. It terminates when then "q" key is touched, for "quit," through the re-setting of the variable `doloop`. Most importantly, it responds to "s" for `stop_servo()` to halt the robot, for "l" to `locate()` it and to "w" for `waypoint_drive()`.

```
int main() {
//int i, j; //for bug test
//for(i = 0;i < 12;i++) for(j = 0;j < 4;j++) dist_stack[i][j] = 99;
screen_init(); //start the screen
open_encoders(); //make sure the encoders are on
display_Eproperties(ENC); //show all the properties first
open_servos(); //make sure the servos are on
display_Sproperties(SRV); //show all the properties first
stop_servo(); //initialize servo positions to neutral
update_display3(); //for testing
receivers_out(); //select receiver locations outside
//receivers_in(); //select receiver locations inside
while (doloop) { //always true until you quit
current_getch = getch();//get a character
```

```
switch (current_getch){ //determine what to do with it
case 97: //'a' for read average
get_mode(); //read encoders, 4 val's returned
break; //terminate case
case 99: //'c' for curve
break; //terminate case
case 100: //'d' for distances
// d_test();
break; //terminate case
case 101: //'e' for encoders
break; //terminate case
case 102: //'f' for feedback
break; //terminate case
case 103: //'g' for go!
servo_go(); //move out
break; //terminate case
case 105: //'i' for NOTHING
break; //terminate case
case 108: //'l' for locate
locate(0,0); //read encoders, 4
break; //terminate case
case 109: //'m' for move
move_servo();//move servos, 2
break; //terminate case
case 112: //'p' for point_list
fileread(); //get a gauntlet on file
break; //terminate case
case 113: //'q' for quit
stop_servo(); //whoa!
doloop = 0; //terminate do loop
break; //terminate case
case 114: //'r' for read file
fileread();
break; //terminate case
case 115: //'s' for stop
stop_servo();//init servos to stop
break; //terminate case
case 116: //'t' for time
GetSystemTime(&time);
millis = time.wMilliseconds;
//secs = time.wSecond;
printw(" %u ", millis); //what time is it??
break; //terminate case
case 118: //'v' for velocity test
// v_test();
break; //terminate case
case 119: //'w' for waypoint
waypoint_drive();//get xd, yd, & speed from list
break; //terminate case
case 120: //'x' for whatever experimental
test_servo();
break; //terminate case
} //this is set up for more cases
}
printw("Closing…\n"); //we quit, so tidy up the house
CPhidget_close((CPhidgetHandle)ENC); //close the phidget
CPhidget_delete((CPhidgetHandle)ENC); //& delete the object
CPhidget_close((CPhidgetHandle)SRV); //close the phidget
CPhidget_delete((CPhidgetHandle)SRV); //& delete the object
screen_end(); //all done, exit
printf("TEST ENDS\n"); //tell the user
return 0;
}
```

## 7.4 History and Diagnostics: The Immediate Past Used for Analysis

These functions save the prior values of many parameters for historical purposes, calculating smoothed data for location or operation, or simply tell the user about the status of the system.

This function loads global variables for display of numbers to the user from the computer to the remote robot. The robot is equipped with 4 bright yellow LEDs that respond to the 16 codes determined by this list.

```
void display() { //use these for the yellow lights
D[0] = 55; //these are values for SRV(2)
D[1] = 59;
D[2] = 65;
D[3] = 71;
D[4] = 78;
D[5] = 85;
D[6] = 91;
D[7] = 97;
D[8] = 104;
D[9] = 110;
D[10] = 117;
D[11] = 122;
D[12] = 130;
D[13] = 135;
D[14] = 142;
D[15] = 149;
}
```

This function is called by interrupt if an error occurs on the Advanced Servo Board #1061. We do nothing but flash the data to the screen for the user.

```
int __stdcall ErrorHandlerS(CPhidgetHandle SRV, void *userptr, int
ErrorCode,\
const char *unknown)
{ //This is an servo board event-triggered call-back function
printf("Servo error handled. %d - %s", ErrorCode, unknown);
return 0;
}
```

This function is also called by interrupt if an error occurs on the Encoder #1047. We do nothing but flash the data to the screen for the user. The Error Code 14 is a special code that crops up occasionally and means nothing, so we ignore it.

```
int __stdcall ErrorHandlerS(CPhidgetHandle ENC, void *userptr, int
ErrorCode, \
const char *unknown)
{ //This is an event-triggered call-back function
if(ErrorCode ==14) return 0; //little trick needed for odd Phidgets
printf("Error handled. %d - %s", ErrorCode, unknown);//standard
response
return 0;
}
```

This function updates the storage of historical distances as found by two receivers onto the `dist_stack[][]`. The formal parameter is the column number of the value to be stored. Here, it is called `stack_num` and can be 0 or 1. Only two values are needed since only two receivers are now used. (It was initially set up to receive 12 distances.) It first moves each value "down" in the list by one row until all are done. This is done the "hard way" rather than using pointers since the processor is fast enough. There can be no stack overflow. The target value, stored as `dist[]`, is placed in the zeroth position, as the most recent. This value is produced by the interrupt routine `InputHandlerE()`. In addition, a second, smoothed value, is placed on a parallel stack, the `smoothed_dist_stack[][]`. This value is stored by the interrupt routine in `avg[stack_num]`. The purpose of the smoothed set of values is to eliminate jitter. As explained earlier, the `InputHandlerE()` passes a line through the past 10 values stored on the stack. The y-intercept of this line is the smoothed value. This algorithm takes out jitter, assuming that it is usually Gaussian.

```
int push(int stack_num) { //vital for storing receiver distance data
int i; //both raw data and smoothed data
for(i = stack_length;i > 0;i--) //move 'em all down one space
dist_stack[i][stack_num] = dist_stack[i-1][stack_num];
dist_stack[0][stack_num] = dist[stack_num];//add top element
for(i = stack_length;i > 0;i--) //move 'em all down one space
smoothed_dist[i][stack_num] = smoothed_dist[i-1][stack_num];
smoothed_dist[0][stack_num] = avg[stack_num];//add top element
return 0;
}
```

This function updates the storage of historical instances of missing distances that are usually found by two receivers. This stack operates the same way the previous one does, but stores only a 0 for "good data" or a 1 for "no data." With this information stored, the user can determine when the values for distances are being extrapolated due to the position of the robot behind a tree. Alternatively, data can be lost if the robot lies close to the line between a pair of receivers, since triangulation becomes impossible. The formal parameter, again, is just the ID of the receiver, either a 0 or a 1, and determines the column number of the value to be stored. The stack is called the `flag_stack[][]`.

```
int pushflag(int stack_num) { //vital for storing receiver range data
int i; //1 = no data
for(i = stack_length;i > 0;i--) //move 'em all down one space
flag_stack[i][stack_num] = flag_stack[i-1][stack_num];
flag_stack[0][stack_num] = flag[stack_num];//add top element
return 0;
}
```

This function updates the storage of historical values as found by the control law running in the tight loop of `waypoint_drive()`. It stores four values onto `stack4[][]`, and the four values are the difference between the current and previous positioning error, `diff`; the commended change to the velocity commands for the drive motors, `delta`; the difference between the current heading and the target heading, `error`; and the current positioning error, `d2line`. The purpose of this stack is to inform the user that the control law is operating normally, that is, it is responding to disturbances caused by uneven turf. There is a limit to the command change that depends on the speed setting, `vel`. This is done to prevent a tight turn that would damage turf. If the `delta` value stays at this maximum value for more than a few cycles, there is a problem with positioning. No formal parameter is needed since only these four values, in order, will be saved. Like the other stacks, this algorithm moves the data rather than a pointer to the data. When done, the data are displayed in the fourth box on the screen by `update_display4()`.

```
int push4() { //used for storing run data
int i; //should match locate's push
for(i = stack_length;i > 0;i--) { //move elements down one space
stack4[i][0] = stack4[i-1][0]; //diff
stack4[i][1] = stack4[i-1][1]; //delta
stack4[i][2] = stack4[i-1][2]; //error
stack4[i][3] = stack4[i-1][3]; } //d2line
stack4[0][0] = diff; //add top elements
stack4[0][1] = delta;
stack4[0][2] = error;
stack4[0][3] = d2line;
screen_init(); //do these when you have something to show
update_display4();
return 0;
}
```

This function updates the storage of historical x, y, and z values as found by the `locate()` function running in the tight loop of `waypoint_drive()`. It stores one value onto the `stack[][]`, depending on the index number, `column`. As many as six values can be saved here, so that differences between the locations and orientations can be examined and analyzed. Any jitter or systemic variations can be spotted in this list, since it is displayed at every cycle in the third data box on the screen. At the start of any run between waypoints that have no previous history, the stack values, and hence the history, will register as zeros. For this reason, there is a function called `startout()` that moves the robot a short distance (about 12 cycles) just to fill the stack with historical data that can be used to remove jitter. Also, since the orientation, `ze`, is computed based on differences between `xe` and `ye` values, it takes several cycles for `ze` to "settle" into a correct pattern of values. These should approximate the target heading, `tar`, as posted on the screen at the start of each waypoint segment. Each of the coordinate values are stored in the variable `coord[]` as generated by the `locate()` function.

```
int pushdown(int column) {//make up a stack of locations for
historical reference
int i; //include distances to target waypoint as well
for(i = stack_length;i > 0;i--) //move elements down one space
stack[i][column] = stack[i-1][column];
stack[0][column] = coord[column]; //add top element
screen_init(); //do these when you have something to show
update_display3();
return 0;
}
```

This function computes the standard deviation of a list of numbers loaded into the column `dex[]`. It can be up to 60 elements long, like the stacks. The function uses a formal parameter, n, to define the length of the list, but it is not checked.

```
float stnd_dev(int n){
float sd = 0;
int i;
for(i = 0;i < n;i++) sd = sd + dex[i];
xbar = sd/n; sd = 0;
for(i = 0;i < n;i++) sd = sd + (dex[i]-xbar)*(dex[i]-xbar);
sd = sqrt(sd/(n-1));
return sd;
}
```

This function initiates and/or appends data to a tab-separated file called "work.txt" for the purposes of later analysis by the user. The data it stores depends on the configuration of the code which may change as the need arises during development. The length of each column of values that it stores is equal to the stack_length-1. This value can be chosen for convenience, but we have set it to 60 so that a single page of data can be listed with a sensible heading. A check is made for a file name, and the file is opened for append. Any error in opening is reported to the screen, and the routine then terminates. Otherwise, after another check for a file name, it prints a heading to the file, followed by 60 values as defined in the `fprintf` statement. A final line is added that gives the average value of the `d2line` column. This value can be interpreted as the average error experienced by the robot during a move along a waypoint line segment during the 60 previous events. The value, `arg1`, is computed by `waypoint_drive()`. This function is called by `waypoint_drive()` after each segment is completed, or the user terminates a move with a "q" and quits.

```
int filewrite(){ //write distance values as we move about
char fname[20] = "work.txt"; //for later quality analysis
int i, j = stack_length-1; //our indexes
//float sdd;
if(fname! = NULL) { //standard recording command
workfile = fopen(fname,"a"); //append
if(workfile==NULL) {printw("Some kind of workfile error.");
return 0;}
}
if(fname! = NULL) { //standard recording command
fprintf(workfile,"\ndiff\tdelta\terror\td2line\txe\tye\tze\td1\td2\td
0\n");
for(i = 0;i < j;i++)
fprintf(workfile,
"%6.3f\t%6.2f\t%6.3f\t%5.1f\t%5.1f\t%5.1f\t%5.2f\t%5.1f\t%5.1f\t%d
%d\n",\
stack4[i][0],stack4[i][1],stack4[i][2],stack4[i][3],stack[i][0],\
stack[i][1],stack[i][2],dist_stack[i][1],dist_stack[i][2],\
flag_stack[i][1],flag_stack[i][2]);
fprintf(workfile,"Avg error: %6.3f \n",arg1);
}
return 0;
}
```

This function initiates and/or appends data to a tab-separated file called "`xyz.txt`" for the purposes of later analysis by the user. The data it stores is fixed to include only those distance measurement values and their consecutive differences. The length of each column of values that it stores is equal to the `stack_length-1`. This value can be chosen for convenience, but we have set it to 60 so that a single page of data can be listed with a sensible heading. A check is made for a file name, and the file is opened for append. Any error in opening is reported to the screen, and the routine then terminates. Otherwise, after another check for a file name, it prints a heading to the file, followed by 60 values as defined in the `fprintf` statement. The `ze` value from the `stack[i][2]` is checked for its principal value. A final line is added that gives the standard deviations of each of the difference values saved in the last three columns. This value can be interpreted as the average jitter experienced by the robot during a move along a waypoint line segment during the 60 previous events. This function is no longer called by the program, but was very useful during development.

```
int xyzwrite(){ //write distance values as we move about
char fname[20] = "xyz.txt"; //for later quality analysis
int i, j; //our index
float dx, dy, dz, sdx, sdy, sdz;
if(fname! = NULL) { //standard recording command
workfile = fopen(fname,"a"); //append
if(workfile==NULL) {printw("Some kind of workfile error.");
return 0;}
}
if(fname! = NULL) { //standard recording command
fprintf(workfile,"\nxe\tye\tze\tdx\tdy\tdz\n");
j = stack_length-2;
for(i = 0;i < j;i++) {
dz = stack[i][2]-stack[i + 1][2];
if(dz > PI) dz = dz-2*(float)PI;
if(dz < -PI) dz = dz + 2*(float)PI;
dx = stack[i][0]-stack[i + 1][0];
dy = stack[i][1]-stack[i + 1][1];
fprintf(workfile,"%6.3f\t%6.3f\t%6.3f\t%6.3f\t%6.3f\t%6.3f\n",\
stack[i][0],stack[i][1],stack[i][2],dx,dy,dz);
}
for(i = 0;i < j;i++) dex[i] = stack[i][0]-stack[i + 1][0];
sdx = stnd_dev(j);
for(i = 0;i < j;i++) dex[i] = stack[i][1]-stack[i + 1][1];
sdy = stnd_dev(j);
for(i = 0;i < j;i++) {
dz = stack[i][2]-stack[i + 1][2];
if(dz > PI) dz = dz-2*(float)PI;
if(dz < -PI) dz = dz + 2*(float)PI;
dex[i] = dz;}
sdz = stnd_dev(j);
fprintf(workfile,"Standard Deviations: %6.3f %6.3f \
%6.3f\n",sdx,sdy,sdz);
}
return 0;
}
```

This function features the standard codes needed to interface with the USB-driven Phidgets boards. We include it here to remove all start-up mysteries. Every line has been checked for good working functionality. You will find it consistent with the web-issued instructions from the supplier, but with no assumptions made.

```
//function to set up USB services for the Phidget encoder board
int open_encoders() {
int i, result; //for an error code, maybe
// CPhidget_DeviceID deviceID; //special type here
const char *err; //just in case, the error message
printw("This version reads the encoders and reports accordingly.\n");
CPhidgetEncoder_create(&ENC); //create the Encoder object;define the
handler
CPhidget_set_OnError_Handler((CPhidgetHandle)ENC, ErrorHandler,
NULL);
CPhidgetEncoder_set_OnInputChange_Handler((CPhidgetEncoderHandle)ENC,
\
InputHandlerE, NULL); //open the interfacekit for device connections
CPhidget_open((CPhidgetHandle)ENC, -1);
printw("Waiting for ENCODERS to be attached...\n");
if((result = CPhidget_waitForAttachment((CPhidgetHandle)ENC,
10000)))//10 seconds
{
CPhidget_getErrorDescription(result, &err); //get error type
printw("Problem waiting for attachment: %s\n", err);//tell user
```

```
what's up
return 0;
}
// if(deviceID == PHIDID_ENCODER_HS_4ENCODER_4INPUT) printw(" OK ");
printw("Enabling Encoders....\n"); //enable each of the four
for(i = 0; i < 4; i++) CPhidgetEncoder_setEnabled (ENC, i, 1);
//CPhidgetEncoder_set_OnInputChange_Handler(ENC, InputChangeHandler,
NULL);\
//define the handler
return 0;
}
```

This function shows the user what occurs during a startup sequence for the Phidget Encoder board #1047. With these lines, one is assured that the connections have been made and that the correct board is "on-line" with your program.

```
//let's see what's on the USB, to be sure it's ours
int display_Eproperties(CPhidgetEncoderHandle phid) { //whose
properties?
int serialNo, version, numEncs;//used by device
const char* ptr; //needed to print messages
CPhidget_getDeviceType((CPhidgetHandle)phid, &ptr); //like this one
printw("%s; ", ptr); //string message
CPhidget_getSerialNumber((CPhidgetHandle)phid, &serialNo);//like this
one
CPhidget_getDeviceVersion((CPhidgetHandle)phid, &version);//and this
one
printw("Serial Number: %10d; Version: %8d; ", serialNo,
version);//see them
CPhidgetEncoder_getEncoderCount(phid, &numEncs); //more about it
printw("# Encoders: %d\n", numEncs); //see it
printw("Press 'q' to quit. "); //reminder
return 0;
}
```

This function features the standard codes needed to interface with the USB-driven Phidgets boards. We include it here to remove all start-up mysteries. Every line has been checked for good working functionality. You will find it consistent with the web-issued instructions from the supplier, but with no assumptions made.

```
int open_servos() { //set up the USB services for the servos
int result, i;
const char *err;
screen_init(); //get the little screen ready
printw("This version reads the keyboard for a command. ");
CPhidgetAdvancedServo_create(&SRV); //create the InterfaceKit object
CPhidget_set_OnError_Handler((CPhidgetHandle)SRV, ErrorHandlerS,
NULL);
//define the handlers
CPhidget_open((CPhidgetHandle)SRV, -1);//open the ifkit for device
connections
printw("Waiting for servo board to be attached…\n");
if((result = CPhidget_waitForAttachment((CPhidgetHandle)SRV,10000)))/
/10
seconds
{
CPhidget_getErrorDescription(result, &err);
printw("Problem waiting for attachment: %s\n", err);
return 0;
}
for (i = 0;i < 4;i++){ //numservos = total of 4
```

```
CPhidgetAdvancedServo_setAcceleration (SRV, i, 500);
CPhidgetAdvancedServo_setVelocityLimit (SRV, i, 5000);
CPhidgetAdvancedServo_setSpeedRampingOn (SRV, i, 1);
CPhidgetAdvancedServo_setEngaged (SRV, i, 1);
} //servos are "on" ; a SAFE position follows immediately
return 0;
}
```

This function shows the user what occurs during a startup sequence for the Phidget Advanced Servo board #1061. With these lines, one is assured that the connections have been made and that the correct board is "on-line" with your program.

```
//Display the properties of the servo phidget to the screen.
//Display the name, serial number and version of the attached device.
//Also display the number of motors on the servo board
int display_Sproperties(CPhidgetAdvancedServoHandle phid) {
int serialNo, version, numMotors;
const char* ptr;
CPhidget_getDeviceType((CPhidgetHandle)phid, &ptr);
printw("%s; ", ptr);
CPhidget_getSerialNumber((CPhidgetHandle)phid, &serialNo);
CPhidget_getDeviceVersion((CPhidgetHandle)phid, &version);
printw("Serial Number: %10d; Version: %8d; ", serialNo, version);
CPhidgetAdvancedServo_getMotorCount(phid, &numMotors);
printw("# of Motors: %d\n ", numMotors);
return 0;
}
```

## Problems

7.1 Suppose time_reads() indicated that a certain function took more than 80 ms to complete. What action could you take to modify the code?

7.2 Suppose curses.h were too slow. How could you modify the system to respond to a single keypress?

7.3 How could the operation of the stacks be made more efficient?

7.4 Why are only three channels displayed by the function update_display_servo(), when four are always used by the LMM3 chip?

7.5 Consider using the display() code to flash the point_count number (modulo 16) on the display board, driven by channel 3 of the *Main Receiver.* Show the new subfunctions needed and where they would fit with the *Inform User* function block.

7.6 Please instantiate the functions of the previous question, as C-code. Be sure to include update_display_servo().

7.7 Consider another approach to smoothing data than those used in the present code, for example, in find_line(). Discuss the entailments of your idea.

7.8 Show with a diagram how triangulate() works.

7.9 Explain how one encoder works to capture the time-of-flight distance measurements of one receiver using InputHandlerE and the USB translator hardware of Figure 5.15.

7.10 Under what conditions would dist2line() fail and cause a run-time error?

7.11 What magnitude of lateral error is allowed if get_line() or get_perp() receives x1==x2?

7.12 Show with a function diagram segment and a flow chart how get_mode() works. Could it be used for every call to InputHanderE?
7.13 Show how find_line() could be made to execute faster.
7.14 Consider Acceleration and Ramping as set in open_servos(). How would a "softer" stop time affect robot operations?
7.15 Does the present "soft start" code affect the robot at any other time?
7.16 Suppose get_jstar() were 30% incorrect. How would that affect robot operation? Hint: refer to Figure 6.36.
7.17 Consider a higher-order (circular or parabolic) approach to extrapolate() and show what the new code would be.
7.18 Under what conditions would startout() create a robot positioning error?
7.19 Speculate on how the lost() function might operate to regain control of an errant robot. Be sure to list your assumptions. Explain your ideas with a new function.
7.20 Please explain the meaning of the last line printed to the work.txt file by filewrite() each time it is invoked by waypoint_drive().
7.21 Could get_heading() be improved? Explain.
7.22 Under what conditions could stnd_dev() fail and give run-time errors?
7.23 In order for open_servos() and open_encoders() to operate properly, what steps must be taken with the linker in Visual Studio Express?

## Note

[1] Lancaster, D. (1996) Active Filter Cookbook, 2nd edn, Newness.

# Appendix A

# Myth and Creativity in Conceptual Design

## A.1 Introduction

Product design is an unstructured but logical problem, for which successive iterations of synthesis and analysis eventually produce approximations to the desired results[1,2]. Our inability to model conceptual design in the form of equations has spawned an array of more qualitative techniques to optimize the parameters of an essentially creative process[3]. Observation of the creative design practice has led to a growing body of research that attempts to model, or at least "capture," what the designer does in the hopes of aiding the efficiency of higher level design processes. Early work has revealed some problems with synthesis techniques actually used by designers in the mechanical domain[4]. The problem with our progress so far is that

> … designers usually pursue a single design concept, and that they will patch and repair their original idea rather than generate new alternatives. This single concept strategy does not conform to the traditional view of what the design process ought to be.

The conceptual phase, from which the "single concept" arises, concerns the problem of coming up with new ideas or new solutions to older problems[5]. Good conceptual design means innovation, and an innovative design comes about when one deliberately tries to create one (e.g.,[6,7]). For example[8],

> An engineer carefully studies power losses in a coal-fired plant and is able to increase efficiency by 0.1%. Another engineer studying the same data conceives of the idea of using direct energy conversion to use the waste heat and increases efficiency by 5–10%.

Although the essential goal (efficiency) is the same, the functions used to produce it are very different. The former approach represents optimization of a given functional model. The literature

*Practical Field Robotics: A Systems Approach*, First Edition. Robert H. Sturges, Jr.

Companion Website: www.wiley.com/go/sturges

abounds with methods for deriving form from such a model and a set of given specifications (e.g.,[9]). The latter approach, however, represents modification of this given functional model. The designer is needed to reach "out of the box" for the chosen approach.

Processes for managing functions in conceptual design and the process of innovating these functions have long been described in the Value Analysis/Value Engineering (VA/VE) literature (e.g.,[10,11,12]). This method, when applied by a small group of engineers is consistently effective in achieving focused creative conceptual design goals, such as attaining a given level of product performance, redesigning to reduce costs below a given threshold, and so on. A study by the American Ordnance Association of a sampling of 2000 of its projects revealed improvements in cycle time, reliability, quality, and maintainability in excess of 60%[13]. That the process works is not debated; product design projects in many corporations in the US employ either an in-house "value team" or consult "certified value specialists." The application of VE is however typically episodic: a design team is brought together for a specific project and promptly disbanded when the specific goals have been achieved. (One may speculate that this irregular use of VE is due to the common practice of organizing design practice into domain-specific specialties.)

The popular TRIZ methodology[14] is itself built upon the key elements of VE. Specifically, "[t]he systematic study of techniques and their functions are the background and foundation of TRIZ." (Emphasis in original.) Unfortunately, TRIZ excludes the breadth required for more abstract conceptual design elements. As Savransky explains, "classical TRIZ does not distinguish information as an independent raw object...." We will see that information is treated in VE by the same techniques as matter and energy, and thus is applicable to software and business method development.

Since VE teaches understanding through functional models, we will model VE with its own methods, and thereby illustrate the value of myth in creative conceptual design. While we will frequently refer to product design, we understand that VE applies equally well to the design of processes. We have employed the terminology of the Society of American Value Engineers in dealing with representations of design practice[15]. Specifically, *function* refers to largely domain-independent characteristics or behaviors of elements or groups of elements of a design. *Allocation* is the process or the result of assigning specifications and resources, and may be domain-specific. The *intent* of a design is expressed by the structure of its functional and aesthetic elements.

## A.2 Analysis of the Value Process

The term "Value Analysis" applies to a disciplined, step-by-step design thinking system, with specific approaches for mind setting, problem setting, and problem solving[16]. It was originally developed for one specific purpose: to efficiently identify unnecessary cost in an existing design. An artifact/system or detailed models of the artifact/system are examined through a formal reverse-engineering process based on the functions of each component. These functions are then mapped to component cost, analyzed, and manipulated to discover design alternatives. The term "Value Engineering" refers to the same set of approaches applied at the conceptual and preliminary stages of design, before detailed models of the artifact/system exist. In this case, the function process is employed as a synthesis tool. We will now show that function analysis/synthesis method of VA/VE can be applied to more creative aspects of design than cost. We will bring in issues not consistently considered in engineering design, most notably, the mythology of a product or process.

### *A.2.1 The Value of Function in Creativity*

A key analysis tool the VA/VE approach is the formal representation of a design in the form of a *function block diagram* (FBD) through the process of *function logic*. Originally developed to stimulate design creativity[17], this non-rigorous but systematic process relies on early identification of design goals and describes them in non-domain-specific functional terms. This element of VA is usually applied to analyzing cost issues, but is not limited to this single interest. In essence, one must think in terms of what the product *does*, rather than what it *is*. In Value Engineering, a designer or team begins with a description of the basic function of the design (Figure a1.1) and a statement of constraints. If the basic function cannot be accomplished by a single component, it is decomposed into several functions that collectively and/or alternatively perform the function. These secondary functions are then translated into components or recursively decomposed. The function decomposition process continues until one can map each function into a component or system that will accomplish it. By system, we mean that the specifications or behaviors of a given function have become sufficiently well characterized that design synthesis tools and methods can be invoked to yield the needed artifacts. The design represented by a list of low-level functions does not directly address the many-to-one problem.

The general form of the FBD is shown in Figure a1.1. The function block (or node) contains a generic function descriptor (what is done) comprised of an active verb and a measurable noun. Alternative definitions of function comprising isolated verbs, nouns, or attributes lack a consistent basis upon which to perform semantic reasoning or logical analysis. For example, if

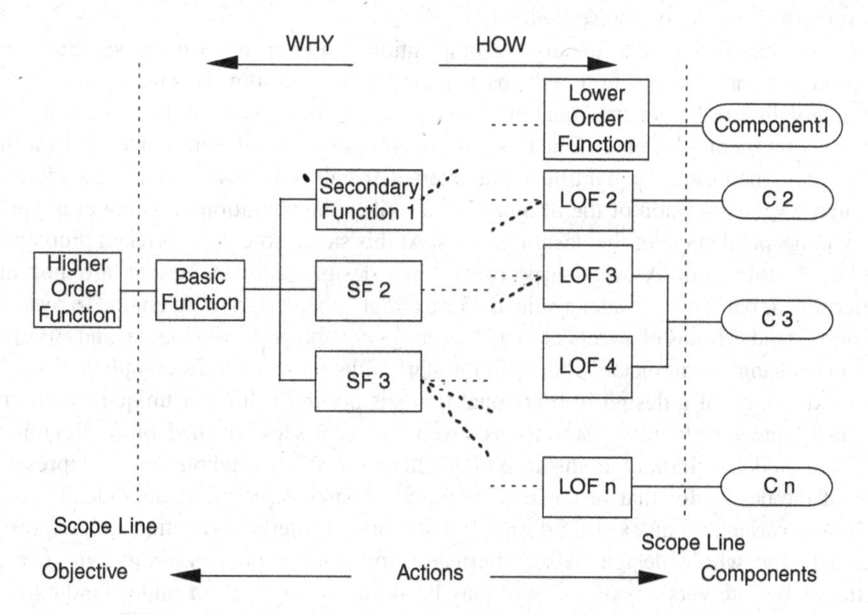


**Figure a1.1** The general form of a FBD. (After Ruggles (1971).[11])

one were to identify "stability" as a function, the context for evaluation or development is absent. Conversely, functions comprising verbs and nouns with modifying adverbs and adjectives unnecessarily constrain or limit the function to pre-existing or assumed conditions. We find creativity practices by non-engineers based on almost identical reasoning. One of the "big seven" lessons learned in extensive creative practice is that "freedom comes from limitations" such as succinctly expressed boundaries[18]. For example, word lists to identify nouns and verbs for application to function descriptors are a common part of the decomposition process.

Texts on the VE approach usually employ a particular subset of classified verbs and nouns[19], namely, nouns which are either qualitative, quantitative, or conceptual, but not concrete. The verb set may include transformation and control, but specifically excludes passive generation verbs such as "allow," "provide," "permit," and so on, as insufficient to describe action or purpose. The nodes to the left of a function node represent the reason *why* a function is included: a higher order function. The nodes to the right are functions describing *how* the function is performed: lower order functions. For example, a high-level function of a mousetrap might be attract mouse; a lower level function just to the right of this might be offer bait. *Why* it offers bait is to attract the mouse. *How* it attracts a mouse is by offering bait. All such identified functions are linked with this how–why relationship in an FBD. Other linkages denoting simultaneity have been proposed[11].

In practice, verb/noun function pairs are written on chits or Post-It™ notes, whence they are rearranged and negotiated by the design team. Computer-assisted implementations of listing, examining, and negotiating are under development (e.g.,[20,21,22]). We prefer to attach another layer of information to each function block to record and manage its allocations[23]. This extension of the standard VA practice leads to better quantitative evaluations of current and possible design states, as will be shown in the examples to follow. In the TRIZ methodology, we find these same elements, namely an active verb represented by the *process* (or TP), the measurable noun by the *subsystem* (or TS), and the allocations by *attributes*[14]. We do not find, however any structure to represent *intent*.

The essential result of the function decomposition process is a reasoning structure relating each product or process function to the basic function of the design. This reasoning structure is deliberately divorced from specifications and constraints that would limit innovation or imply a particular problem solution. In fact, this stage of analysis invites the introduction of conflicting sets of requirements and unquantified parameters in the allocations attached to each function[24]. This high-level expression of the design also describes the conditions existing at the preliminary or conceptual stage of the design process. At this stage, so-called "wicked problems" are posed and addressed. A wicked problem[25] has design parameters that are not merely undetermined, but are also indeterminate. For example, the design of an engine to meet future emissions standards has elements of "wickedness" since the political climate, and EPA regulations, may change unpredictably between the start of the project and its completion.

The expression of a design in functional terms is no less valid, nor unique, when given a new set of indeterminate constraints as given last year's established ones. Retaining the functional model, in light of an instance of an allocation set, is valuable since it represents the intent of the design for that instance. This function-based structure is also ideally suited to developing variants in an existing design[26]: additional or alternate functions are expressed in relation to the whole design. Also, alternate allocations express variants within a given function. Obsolete versions of a design may be similarly retained for understanding product history[23].

Perhaps the greatest value of the functional representation lies in the management of innovation. Manipulation of the FBD creates alternative designs with a deep kinship to the original, but for which new allocation opportunities arise. It is when we determine the allocations attached to each function, that is, quantify the limits of the design space[27], that we attempt to take the "wickedness" out of a problem. With a set of established constraints, linked to the context of function, design synthesis and optimization can often proceed to a finite set of alternative solutions[28]. As the problem parameters shift due to time and technology, we negotiate new allocations without changing the functional essence of the design.

## A.2.2 *The Function of Value in Creativity*

It is tempting to employ functional analysis on the practice of design itself. Models of engineering design practice derived from protocol studies indicate a diversity of views on design data[21] and process[29]. However, the evidence shows that almost all such activity occurs to the *right* of the rightmost scope line of Figure a1.1. Designers tend to work from interpretations of the lowest level functions expressed by their allocations. At this level, fundamental errors may be missed; for instance, highly advanced on-screen television programming systems have been designed and marketed with apparently little concern over whether the average person will ever learn to use them. A broader view of design is needed to show the higher level functional concerns and how they are expressed in conceptual and preliminary designs.

The model proposed by Buchanan[30,31] more explicitly states the basic functions in conceptual design and the form of their allocations, namely ideas about products, the internal operational logic of products, and the desire to use products. Expressed in FBD form (Figure a1.2), we find the basic function, Design Product, decomposed into three subfunctions, all of which must be present to assure product integrity. Of course, one may substitute Process for Product here with no loss of meaning. In this way issues of concurrency in product/process design become memorialized. Recall that an FBD is not a flow chart, but rather a statement of activities in a process connected by reasoning which supports their existence. In Figure a1.2, the verb Express is understood to be active: to realize, demonstrate, or manifest its object.

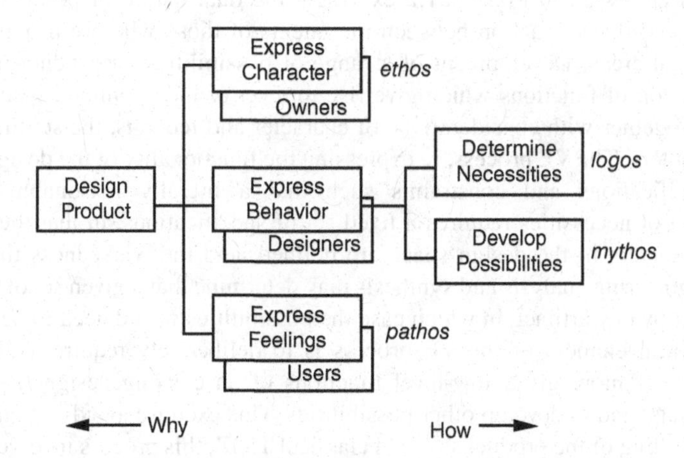


**Figure a1.2** An FBD representation of design practice. (After Buchanan (1985).[30])

Following Buchanan, ideas about the product, in Express Character, are needed to articulate the overall quality, or *ethos*, of the product and is usually allocated to the process owners, or makers. The "voice of the product," the desired aesthetic is explicitly defined by its *ethos*. For example, it expresses the differences between a plastic fork and a fine piece of sterling, or a personalized coffee mug and a styrofoam cup. The desire to use a product, in Express Feelings, is needed to create, anticipate, and measure the user reaction, or *pathos*, to the product. For example, it expresses the differences in emotional experience between driving a sports car and a taxi cab. At this level, the failure of "New Coke™" in the marketplace seems less mysterious, since the "old Coke" had a life of its own in the feelings of its users.

The major activities in engineering and industrial design are found in the Express Behavior function, in which the term behavior applies to the internal operational logic of a product: the physical expression of the artifact and its interactions with its user and its environment. At this level of expression, we recognize two subfunctions. Determine Necessities expresses what the product is in terms of materials, configuration, and physical interfaces. Systematic design methodologies (e.g.,[32]) view the design process as linear with conceptual design as the first stage. Pahl and Beitz[33] characterize the conceptual stage by identification of the essential problem, the determination of function structures, and the search for solution principles. This process of satisfying physical laws and practical constraints is carried out by rigorous reasoning, or *logos*, in a cycle of analysis and synthesis usually allocated to the engineering designer. The other subfunction, Develop Possibilities, expresses what the product does in the user context. This process of satisfying functional desires and constraints is carried out through a "true narration," or *mythos*, in a cycle of ideation (forming in the imagination) and realization usually allocated to the industrial designer. The formal expression of product mythology can become a focus of high level conceptual design activity. Collaboration between these functions of *logos* and *mythos* is evidenced, for example, in the Black & Decker Dustbuster™, which deliberately combines the (necessary) functionality of the vacuum cleaner with certain (possible) functional aspects of the telephone. Such creative practice is found in current methodologies. Specifically, we find that no less than 11 of the 52 creativity exercises recently expressed by a non-engineer[18] parallel the VE process, including story-telling, inversion, experimentation, and charting methods.

The model of design in Figure a1.2 expresses the dual nature of product specification development and the interaction between the talents of those who are usually assigned to these functional areas. Development of a range of possibilities, or product myth, requires conceptualization of functions which give rise to a set of indeterminate specifications and constraints. Together with considerations of character and feelings, the set may comprise a "wicked" problem. The VE process, in expressing the functionality of the design, attempts to allocate specifications and constraints such that a literally reasonable set remains. Determination of necessities requires a fixed set of specifications, in that the relationships between parameters in the design space are defined, and the wickedness thus apparently resolved. Engineering analysis and synthesis may determine that a given set of specifications cannot be met by any artifact, in which case the possibilities would need to be re-examined. Indeed, a central element of the VE process is to deliberately require a design team to disqualify one or more given low-level functions of an existing design (and all of their implied artifacts) and to develop other possibilities. This exercise avoids repeating old ideas through a re-telling of the product story. In classical TRIZ, this process involves an ontology of *contradictions*[14].

Wholly new products benefit from the domain-free expression of myth and logic as in the case of software authorship. Here, the intent of creative design is to satisfy functional requirements that are typically expressed in a sequential fashion that is not conducive to innovative thinking about design, namely the flow chart. Most typically, the structure of software is strongly influenced by this linear approach to communicating ideas to colleagues, when the option of functional expression has been well-developed but not applied to this field. While counter-intuitive, the Value Engineering approach applies a multistep, or "waterfall" approach to solving systemic problems in a product or process. Software is commonly managed by such a linear process, but the differences are not at all subtle: the requirements specification in VE is expressed in language (the FBD) that all stake-holders give approval before the coding begins. Furthermore, the inclusion of *mythos*, or "what could it do?," is tied directly to the "how" and "why" of the product intent.

A brief review of coding methods (e.g.,[34]) reveals that no software product development processes such as Agile, Spiral, Iterative, or even "Cowboy" coding methods capture, let alone begin, with intent. We will show in the next section examples of a method more conducive to creative coding.

In short, the VE process maintains the tension between creative concept ideation and realization by expressing both at a domain-free functional level. The FBD also serves to develop specifications through the processes of function identification and allocation, such that the designer can proceed with more detailed engineering analysis and synthesis. Restricting VE to the limited purpose of cost reduction misses the opportunity to innovate function through development of a product *mythos*.

## A.3 Examples

Insights derived from functional modeling of aspects other than technical performance can be seen from the next three examples. In the first, identification of a *mythos* for a chair predominates the creation of variant designs and memorializes an understanding of the relationship of the user to the product. In the second, a new concept in wearable computers was taken from its initial prototype form to its second model by reverse engineering its internal operational logic. The extended FBD with allocations served to re-specify the design and focus subsequent design efforts. Finally, synthesis of a new product is presented: a device for evaluating assembly effort. In this case, functional modeling of all design aspects was considered at the conceptual and preliminary phases.

### *A.3.1* A Mythos *for Chair Design*

To illustrate the function of VE and its relationship to expressing the internal operational logic of a product, let us examine a familiar item which enjoys a continuing redesign process[35,36]. The conceptual design of a chair is partially expressed in Figure a1.3 using the model of Figure a1.2 and the FBD language. We have restricted our scope to the Express Behavior subfunction, therefore leaving out the logic which gives rise to specifications of aesthetics, weight, cost, intended point of use, and customer. Recall that any of these allocations may be indeterminate and that the allocation process is a parallel design activity.

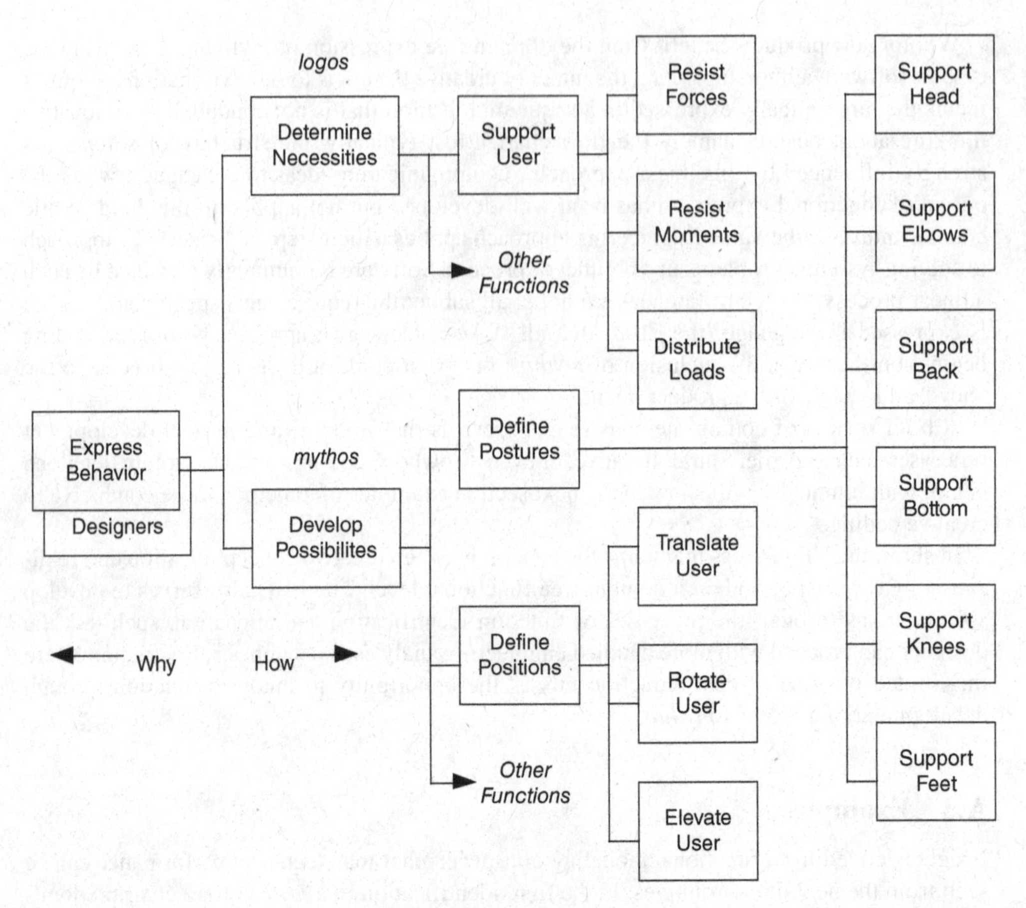


**Figure a1.3** Partial FBD of a chair

The necessary functions of the chair point to configurations of structural elements such as legs, back, and seat which interact to satisfy the functions Resist Forces and Resist Moments. A third necessary function, Distribute Loads, expresses the chair/user interface at the points of shared pressure. Each of these three functions (and there may be others) would be further decomposed by a designer until a collection of artifacts, constraint equations, or deterministic methodologies evolves which satisfy the allocations inherited by each lower level function from its parent(s). The *logos* of the chair is then complete. At this point the form is largely determined by the designer's interpretation of what a chair ought to be, or perhaps, what last year's model looked like. Innovation is a subjective matter here, and indeterminate allocations will be decided by choice. The FBD (and attached allocations) explicitly expresses the context in which these subjective judgments are made. Failing to note these choices explicitly, taken or not taken, represents a loss of potentially valuable historic design information. This step is wholly absent from contemporary design methodology. Further, the practice of intentionally barring a past choice so that a previously unconsidered technology, for example, is selected leads the designer to innovation by combination of other work in new ways.

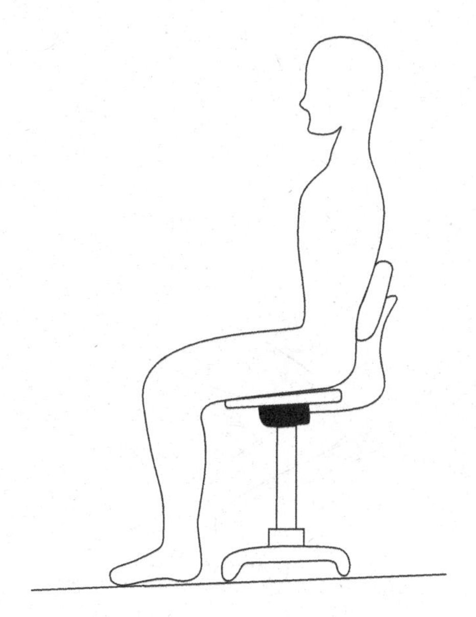

**Figure a1.4** A conventional chair

The possible functions of the chair express either old behavior which apparently satisfies no requirements or, better, new behavior which does not already exist. New functional behavior is developed by the designer in response to allocations from higher level functions, or from newly acquired understanding and appreciation of the present artifact. For example, an essential relationship between the user and chair is explicitly expressible as the "creation of postures." Selecting which parts of the user's anatomy is to be supported produces high level distinctions between different chair designs. Choosing to support the bottom and back of the user leads to a familiar class of designs (Figure a1.4), but choosing instead the bottom and knees of the user leads to an innovative class of chair designs (at least at the time of its introduction) as in Figure a1.5. One can readily imagine the *mythos* of chairs ranging from bar stools to recliners by selecting postures suggested by support points on the user's anatomy. What the chair *could be* is expressed in functional form with a more generalizable model that the non-engineer can apprehend and improve upon in a realizable form[18].

Similarly, another essential relationship between the user and intended use of the chair is explicitly expressible through the chair's function as a positioning agent. Selecting which user motions are aided or restrained produces high-level distinctions between alternative designs. For example, piano benches give lateral sliding motion with no change in elevation, while barber's chairs offer easy rotation and adjustable height while constraining all other motions. It is important to realize here, that the creation of a product *mythos* is not merely to develop a taxonomy of existing chairs, but to develop *new* ones based on the higher level understanding of the product as expressed by its function logic. Lest one conclude that VE reduces innovation to a mere combinatorial exercise, consider that the act of expressing the possibilities, the *mythos*, remains a creative and largely social task. Once created, the *mythos* exists independent of the words used to express it. The list of *other possibilities* in Figure a1.3 is fortunately unlimited and open to the creation of chairs which function in new ways.

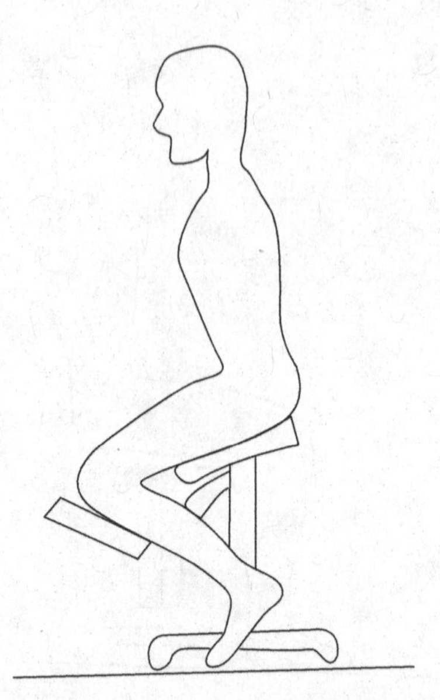

**Figure a1.5** An innovative chair

At some point, the expressed behavior of the product is tentatively determined in sufficient detail to consider moving some *possible* functions to the *necessary* branches of the FBD for more detailed analysis and synthesis. The designer and/or team suspends analysis of the indeterminate allocations and sets values for them anyway. This is the point at which the specifications originate and the wickedness of the problem is ignored. The popular TRIZ method does not, in contrast, suggest such crossover, but rather deals with the contradictions and constraints separately from the non-domain-specific functional models.

## *A.3.2 Functional Variants through Reverse Engineering*

As a second example, we consider the redesign of a wearable personal computer developed in our laboratory[37]. The initial model "VuMan" computer was developed to convey information to a field user with a "head-up display." An FBD was developed by reverse engineering prior to the design of the second model (Figure a1.6). The FBD was developed to highlight the problem areas rather than the entire design. For example, functional descriptions of software were not considered, although the value of such analysis is becoming appreciated[38]. (Indeed, software design using the FBD is now routine in our design classes.) Problems with the first model, in both functions and allocations, were recorded and used to transform the FBD into a new conceptual model (Table a1.1). The first and the new models shared the same FBD to the left of the troublesome functions of the first. Cost was only one of many allocations considered[39]. The expressed *mythos* of the VuMan highlighted critical relationships between many features, for example, a spatial mouse and virtual footsteps for user navigation, voice recognition and synthesis for interactive communication, and allocations expressing convenience for right- and left-handed users. In addition, the design team grew the FBD to describe a realistic VuMan concept for computer-assisted VE. The *logos* of the second model, as produced, is derived from its own *mythos*, which remains as the basis for future innovations.

### *A.3.3 Synthesis of a Design Tool*

In this example, we describe the synthesis of a "new" product in the sense that a product developed from existing elements can be considered new. The product arose from an expressed need for an alternative "design for assembly" (DfA) evaluation tool, where many others existed in the marketplace. The product development is traceable in concept (although not in chronology) from its extended FBD shown in Figure a1.7.

The *ethos* of the product was established early by the cognizant manager and expressed by the conceptual designer with only two identifiable functions: Project Quality and Project Neatness,

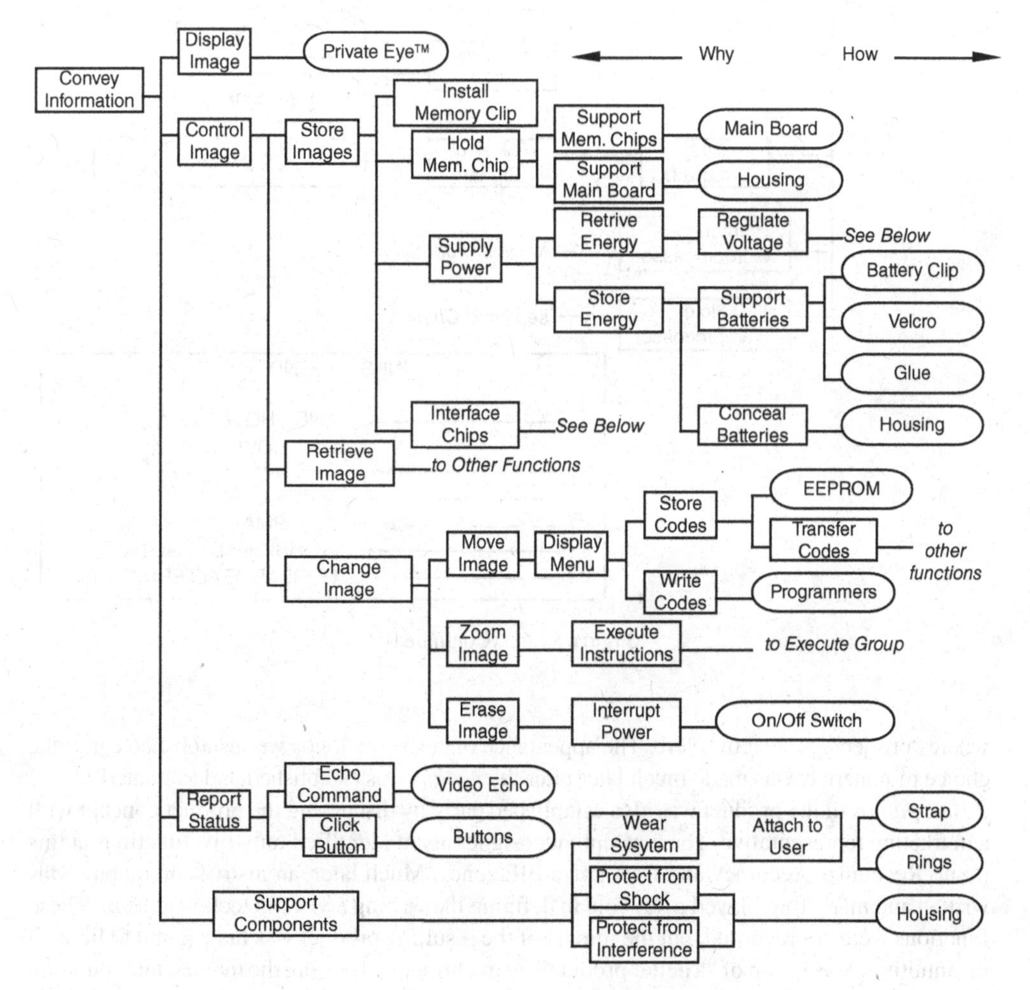


**Figure a1.6** FBD of the VuMan wearable computer: first mode

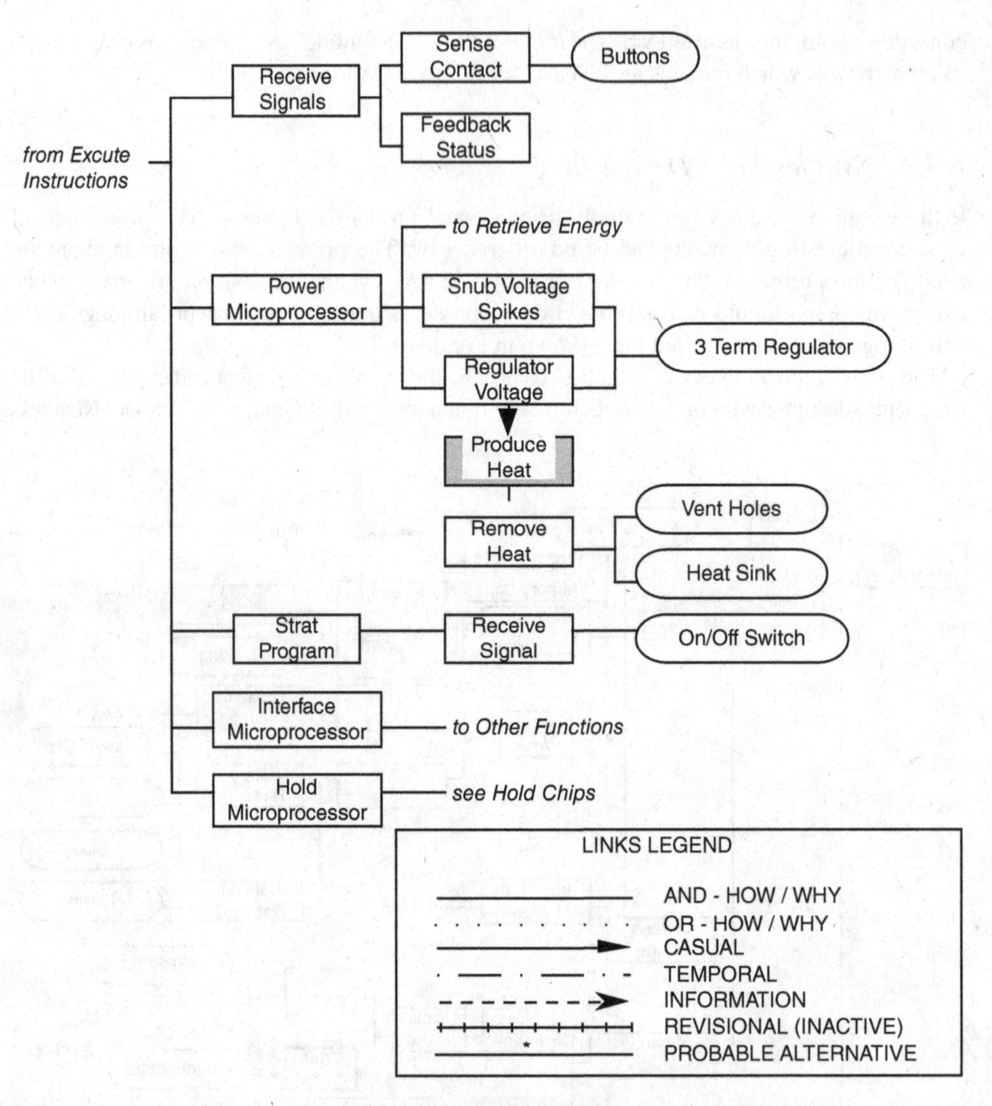


**Figure a1.6** (Continued)

where "pro-jéct" is an active verb. The appearance of corporate logos was established early; the choice of materials was made much later after the *mythos* was established and evaluated.

The *pathos* of the product was also established early by discussing the product concept with a marketing representative. The conceptual designer again identified only two functions at this point: Recognize Accuracy and Recognize Efficiency. Much later, an instruction manual was written and marketing played a key role in defining the writing style and technical level. These functions were not recorded, but the impact of the resulting product was later found to lie with the intuitive expectation of "a better product" by intentionally limiting the manual manipulation of the tool itself.

**Table a1.1** Error-related functions and allocations for the VuMan wearable computer

| Design flaw | Related function |
|---|---|
| Battery compartment lid will not open | Replace batteries |
| Reset button omitted | Erase image |
| Overheating components | Produce heat (side effect) |
| Poor portability while programming | Encode instructions |
| Excess wiring for switches | Interface |
| | Microprocessor |
| Difficulty in loading/changing data base | Store image |
| **Allocation flaw** | **Related allocation** |
| Heat sink too wide | Size |
| Reset button pins touch EEPROM | Size |
| Excess heat in housing | Supply voltage, power budget |

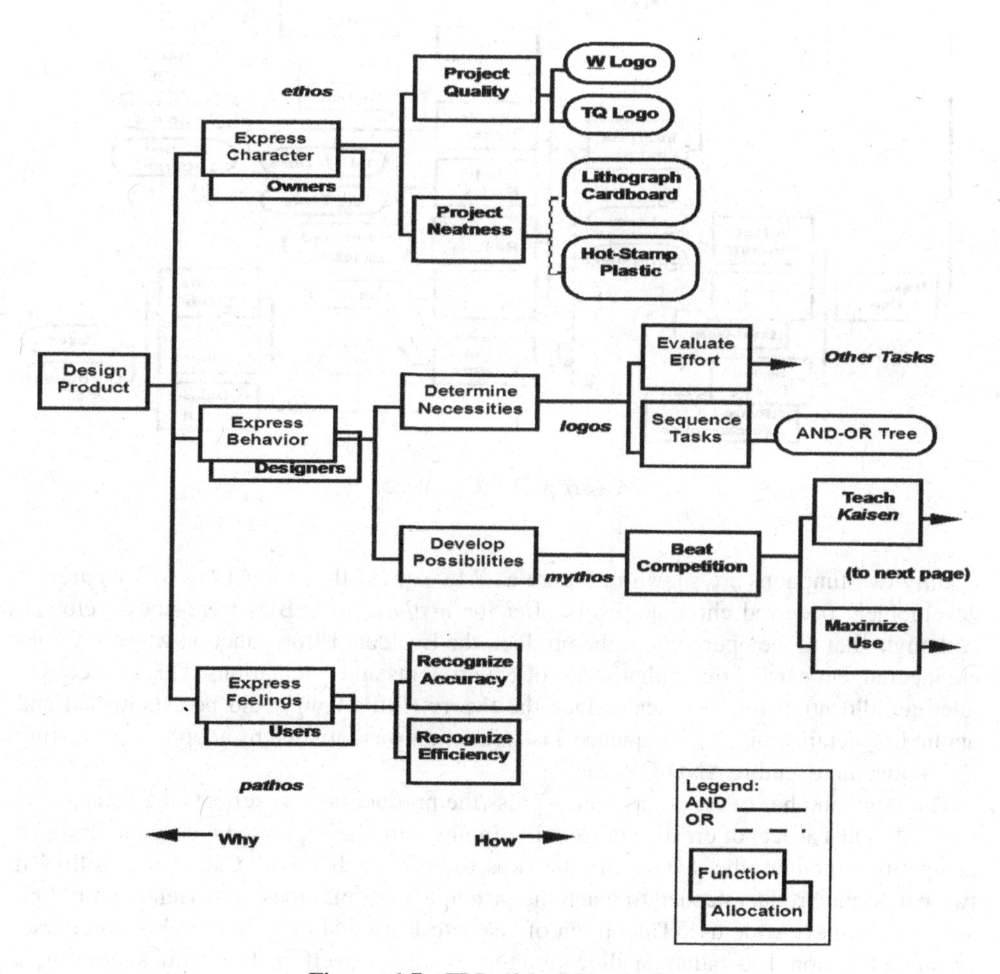


**Figure a1.7** FBD of the DfA calculator

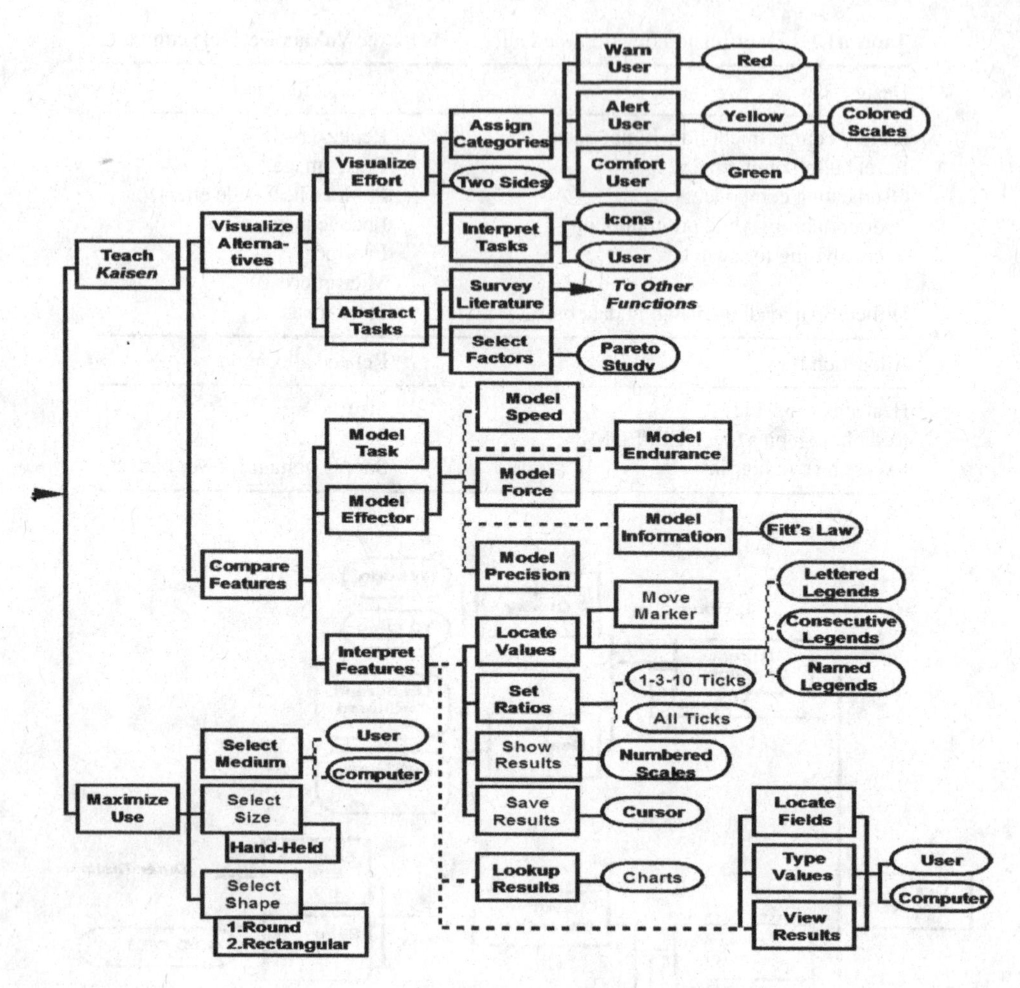


**Figure a1.7** (Continued)

Only two functions are shown in Figure a1.7 to express the product *logos*. This area of development occurred chronologically after the *mythos*, and FBDs were not developed. Although vital to the operation of the product, the Evaluate Effort function was not value-engineered, but carried out with a series of experiments and calculations. This step constituted an allocation for each scale since the theory (Fitt's law) could be established and applied, as detailed in[40]. The Sequence Tasks function was satisfied by adopting an existing technique: an assembly AND-OR tree.

The large number of functions that express the product *mythos* reflects the actual time devoted to this aspect of creative design. Beginning with the expressed mission to design a competitive product, the *mythos* of what was to become the "DfA Calculator" followed two paths: the product needed to teach the principle of continuous improvement, or "kaisen," and achieve wide use. The notion of "easy to learn and use" is desirable, of course, but not a function. It is rather an allocation that expresses itself in all functions requiring a given level of skill or memory by the user. The function block Maximize Use represents a strategic decision, since its converse (Restrict Use, say, to a given domain or market) is an alternative. Its decomposition led to several key allocations, most importantly, the choice of either a computer-based or a manual tool. The dotted lines connecting functions or artifacts indicate an explicit OR relationship. The choice of operating mode was kept open during its preliminary design and development. In fact, the first product was a manual tool in the form of a two-sided circular slide rule. Later, a computer-based version was developed, one screen of which is reproduced in Figure a1.8, but it lacked popularity due to the non-intuitive nature of a keyboard and screen which are static by design, and only representative of motion.

A complete discussion of the balance of Figure a1.7 is beyond the scope of this book, but a few points should be made to emphasize the importance of a fully expressed mythology in this case. A pivotal decision regarding the choice of DfA methodology conveyed by the product (in a sense, what the product actually delivers to the user) is expressed by the Model Task and Model Effector functions. When these functions are taken together, rather than separately, a novel approach to assembly analysis results[40]. The implementation details of this novel approach are expressed by the OR relationship among many candidate metrics. One of them, Model Information, eventually pointed to Fitt's law and the expansion of the Evaluate Effort function, mentioned above.

The myth of the DfA Calculator demands essential visual and haptic relationships between user and tool. As the user Locates Values and Sets Ratios by Moving a Marker, a sense of effort is conveyed by the distances traversed. This apparently subtle hand–eye task was found, in subsequent interviews, to create a strong feeling of awareness in the user that certain features were more troublesome than others. This awareness is absent in previous tools. Similarly, the visualization of effort through icons and colored scales expressed the *mythos* of the product long before its form was determined.

### *A.3.4 Application to Software Development*

The amount of coding actually done to realize the foregoing example is too voluminous to be addressed here, so we will take a more tractable example from recent developments in our teaching lab. An example FBD is shown in Figure a1.9, wherein the highest level function is placed to the left, and its decomposed lower level functions appear more rightward. Notice that the "flow" of the knowledge is expressed by answering the question "why" by looking towards the left, and the question "how" by looking towards the right.

The functions themselves are expressed by a single active verb and a measurable generic noun in a small rectangle. For example, "interface display" is a viable function, whereas "provide interface" is not[41]. There is no "right or wrong" function diagram, only viable and non-viable ones as judged by the team of users that employ them for interdisciplinary communication purposes. Please notice also that such a diagram is not intended to express the same information as a flow chart since the latter rarely explains the *raison d'être* for the functions and their logical relationships. In our example, the right-most series of functions, when read from the top to bottom concisely express the "pseudo code" prevalent in other work. Taken all together, the function diagram expresses the intent of the designer or user. The decision to employ a certain platform and language, for example,

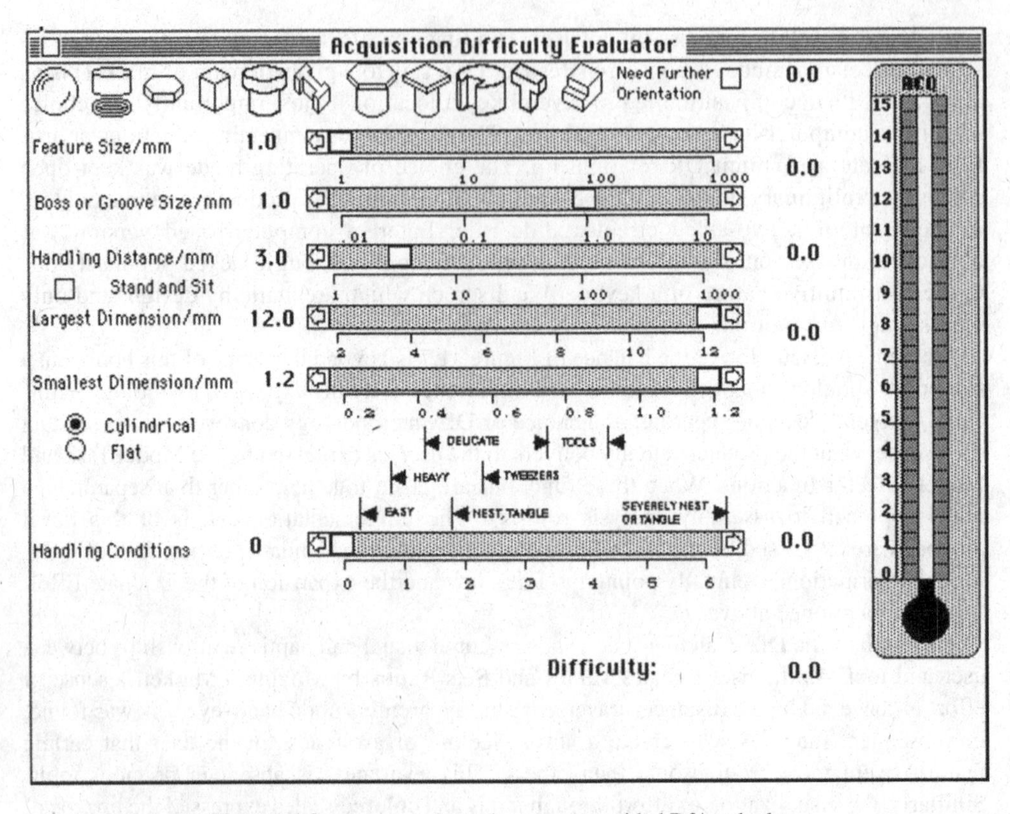


**Figure a1.8** A screen from the computer-aided DfA calculator

Windows 32 and C#, is suspended for later evaluation or recommendation for reasons that may be unconnected with the functions of the code itself.

## *A.3.5 Application to Business*

As a final example, we have found that the FBD aids in finding new pathways for doing the routine or "policy-directed" activities of the office. Where "policy" directs certain information processing activities, a reverse-engineering approach to analyzing the existence and purpose of business activities has proven useful. Not only does it bring all steps of a process into visibility, but it directs the analyst to answer the oft-missing information related to *why* some processes exist at all. The opportunity for myth creation (and "myth-busting") is high, since each function block can be systematically challenged for relevance to the new environment in which it now exists and functions. The *intent* of the policy when it was originally created becomes visible to those responsible for its execution[42].

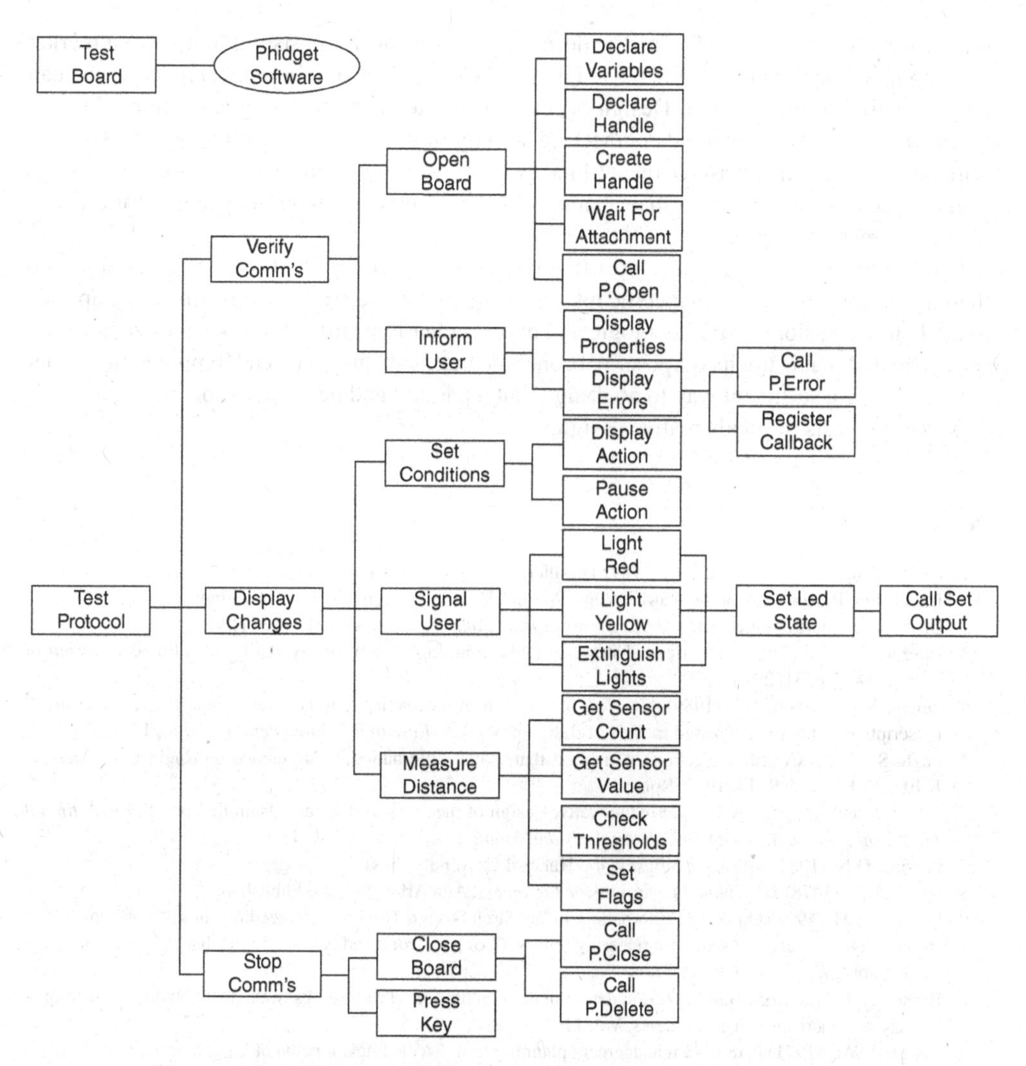


**Figure a1.9** FBD expressing the beginning of a successful software project

## A.4 Discussion and Conclusion

The functional model of design practice as expressed in Figure a1.2 and in the examples helps explain the use, misuse, and non-use of VE. Innovation evidently occurs whether or not a designer is conscious of the process by which he/she explores new possibilities. Function logic may not explain the personal thought processes of the creative designer, indeed the episodic application of VE indicates that such is an unwelcome burden for many. When applied as a discipline, VE contributes to innovation by satisfying functional desires and constraints (*logos*) while expressing concept ideation and realization (*mythos*) at a domain-free level.

When one restricts Develop Possibilities to cost allocation, one gets a new design at lower cost with no planned innovation. This misuse may be good or bad depending on the larger issues of *ethos* and *pathos*. The maker designing only for lower costs will not serve a market desiring new functionalities. When one ignores the development of a product *mythos*, one can only optimize a given design. Design becomes restricted to negotiating allocations. In both cases there is a large, structured engineering activity (*logos*), and the result may or may not be improved design. The measures one ordinarily employs to determine design improvement, for example, cost, structural performance, and so on, may indicate an optimal configuration for a dwindling market.

On the other hand, there can be no artifact without the translation of low level functions and their allocations into domain-specific instances and realizable artifacts. Function logic applied to explicit expressions of product logic and myth is clearly insufficient to synthesize and analyze form, but that is not its purpose. We conclude that such an expanded VE process is neither a different nor a sufficient way to do design, but a parallel and necessary process for the analysis and synthesis of good creative design.

## Notes


1. Paz-Soldan, J.P. and Rinderle, J.R. (1989) The alternate use of abstraction and refinement in conceptual mechanical design. Proceedings of the ASME Winter Annual Meeting, December 1989, San Francisco, CA.
2. Ulrich, K.T. and Eppinger, S.D. (1995) *Product Design and Development*, McGraw Hill.
3. Havenhand, L. K. (2006) American abstract art and the interior design of ray and Charles Eames. *Journal of Interior Design*, **31**, 29–42.
4. Finger, S and Dixon, J.R. (1989) A review of research in mechanical engineering design. Part I: descriptive, prescriptive and computer-based models of design processes. *Research in Engineering Design*, **1**, 51–67.
5. Pugh, S. (1981) Concept selection—a method that works. International Conference on Engineering Design, ICED 1981, March 9–13, 1981, Rome.
6. Cagan, J. and Agogino, A.M. (1987) Innovative design of mechanical structures from first principles. *Artificial Intelligence for Engineering Design, Analysis and Manufacturing*, **1**(3), 169–189.
7. Perkins, D.N. (1981) *The Mind's Best Work*, Harvard University Press.
8. Bailey, R.L. (1978) *Disciplined Creativity for Engineers*, Ann Arbor Science Publishers.
9. Koppelaar, H. (1992) On design automation, in Research Design Thinking. Proceedings of a Workshop Held at the Faculty of Industrial Design Engineering (eds N. Cross, K. Dorst and N.F.M. Roozenburg), Delft University of Technology, Delft University Press.
10. Bytheway, C.W. (1965) Basic function determination techniques. Proceedings of the Fifth National Meeting—Society of American Value Engineers, Vol. **11**, April 21–23, 1965.
11. Ruggles, W.F. (1971) FAST—a management planning tool. SAVE Encyclopedia of Value, Vol. **6**, p. 301.
12. Miles, L.D. (1982) *Techniques of Value Analysis*, 2nd edn, McGraw Hill Book Company, New York.
13. Prendergast, J.F. and Westinghouse Corporate Value Analysis Staff. (1982) *Value Analysis Handbook*, Westinghouse Productivity and Quality Center.
14. Savransky, S. (2000) *Engineering of Creativity*, CRC Press.
15. Sturges, R. H., O'Shaughnessy, K. and Kilani, M. I. (1990) *Representation of Aircraft Design Data for Supportability, Operability, and Producibility Evaluations*, Carnegie Mellon University Engineering Design Research Center.
16. Crow, K. (2002) *Value Analysis and Function Analysis System Technique*, DRM Associates.
17. Bytheway, C.W. (1971) The creative aspects of FAST diagramming. Proceedings of the SAVE Conference.
18. Scalin, N. (2011) *Unstuck*, Voyageur Press.
19. Jakobsen, K., Sigurjónsson, J. and Jakobsen, Ø. (1991) Formalized specifications of functional requirements. *Design Studies*, **12**(4), 221–224.
20. Prentiss, T. (1992) *A PC-Based Value Engineering Tool*, Xerox Corporation.
21. Subrahmanian, E., Podnar, G. and Westerberg, A. (1989) *n-DIM: n-Dimensional Information Modeling—A Shared Computational Environment for Design*, Carnegie Mellon University Engineering Design Research Center.

22. Sturges, R.H. and Kilani, M.I. (1992) Towards an integrated design for an assembly evaluation and reasoning system. *Journal of Computer Aided Design*, **24**(2), 67–79.
23. O'Shaughnessy, K. and Sturges, R.H. (1992) A systematic approach to conceptual design. ASME Fourth International Conference on Design Theory and Methodology, September 1992, Phoenix, AZ.
24. Reed, R.G. and Sturges, R.H. (1993) Toward performance-intelligent design advisors. ASME DAC, September 19–22, 1993, Albuquerque, NM.
25. Rittel, H.W.J. (1984) Second-generation design methods, in *Developments in Design Methodology* (ed. N. Cross), John Wiley & Sons, Ltd, pp. 317–327.
26. Hundal, M.S. (1991) Use of functional variants in product development. Proceedings of the ASME Design Theory and Methodology Conference, September 22–25, 1991, Miami, FL.
27. Kantowitz, B.H. and Sorkin, R.D. (1987) Allocation of functions, in *Handbook of Human Factors* (ed. G.I. Salvendy), John Wiley & Sons, Ltd, pp. 355–369.
28. Westerberg, A., Grossmann, I., Talukdar, S., *et al.* (1989) *Applications of Artificial Intelligence in Design Research at Carnegie Mellon University's EDRC*, Carnegie Mellon University Engineering Design Research Center.
29. Ullman, D.G., Dietterich, T.G. and Stauffer, L.A. (1988) A model of the mechanical design process based on empirical data. *Artificial Intelligence for Engineering Design, Analysis and Manufacturing*, **2**(1), 33–52.
30. Buchanan, R. (1985) Declaration by design. *Design Issues*, **2**(1), 4–22.
31. Buchanan, R. (2001) Design and the new rhetoric: productive arts in the philosophy of culture. *Philosophy and Rhetoric*, **34**(3), 183–206.
32. Pugh, S.. (1991) *Total Design—Integrated Methods for Successful Product Engineering*, Addison-Wesley.
33. Pahl, G. and Beitz, W. (1992) *Engineering Design: A Systematic Approach*, Springer-Verlag.
34. Beck, K. (2000) *Extreme Programming*, Addison-Wesley.
35. en.wikipedia.org/wiki/Eames_Lounge_Chair.
36. Eidelberg, M., Hine, T. and Kirkham, P. (2006) *The Eames Lounge Chair: An Icon of Modern Design*, Merrell Holberton.
37. Siewiorek, D.P., *et al.* (1992) Vu-Man: A Wearable Computer. Carnegie Mellon University Engineering Design Research Center.
38. Kellner, M.I. (1989) Software process modeling: value and experience. SEI technical review, Carnegie Mellon University, pp. 23–54.
39. Sturges, R.H. and Reed, R.G. (1993) An application of extended function logic to the design of a wearable computer. 9th International Conference on Engineering Design, August 17–19, 1993, The Hague.
40. Sturges, R.H. (1989) A quantification of manual dexterity: the design for assembly calculator. *Journal of Robotics and Computer Integrated Manufacturing*, **6**(3), 237–252.
41. Sturges, R.H., O'Shaughnessy, K. and Reed R.G. (1993). A systematic approach to conceptual design based on function logic. *International Journal of Concurrent Engineering: Research and Applications*, **1** (2), 93–106.
42. Leubbe, A. and Weske, M. (2012) Determining the effect of tangible business process modeling, in *Design Thinking Research* (eds H. Plattner, C. Meinel and L. Leifer), Springer-Verlag, pp. 241–257.

# Appendix B

# Real-World Automation Control through the USB Interface*

## B.1 Introduction

The days of ubiquitous memory-addressable I/O are long gone (1950–1985), but the last vestiges disappeared only recently with the phasing out of the printer port, and the adoption of the "Universal Serial Bus," or USB. Thus, a large class of computer peripherals and related software were obsoleted long before they were functionally unable to compete. (A large number of "converters" were developed to fill this gap in functionality.) While printer ports existed one could very easily turn an external light on and off by simple programming[1]. In this appendix, we will review the current needs to do this simple task. We will find that the newer protocols have added additional layers of software and firmware between the programmer and the real-world application.

Let us clarify the meaning of the term "real-world" in this instance. The intensely market-driven world of personal computers (PCs) has led to the seamless integration of multimedia with business support software and interfaces limited to keyboards, mice, and glowing rectangles. As this appendix is drafted, we are seeing the replacement of keyboards and mice with touch-screen interfaces for the consumer who wants to manipulate pre-defined graphical images. Control or interactivity with anything else, such as a simple light bulb or non-standard keyboard is practically out of the question if it does not support a large consumer market segment. This "virtual world" by contrast, is pre-packaged and pre-defined for the consumer based on extensive analysis of taste and desires of a public conditioned to consider the computer as an appliance rather than a device for the implementation of specific logical needs[2]. The boundaries of "being digital" today are not surprising since the vast majority of installed computers are operationally dependent on a monopoly.

---

*This appendix was originally published as: Robert H. Sturges, "Real-World Automation Control through the USB Interface", *International Journal of Engineering Inventions*, Volume 2, Issue 6, April 2013. Reproduced with permission from *International Journal of Engineering Inventions*. Copyright 2013.

*Practical Field Robotics: A Systems Approach*, First Edition. Robert H. Sturges, Jr.

Companion Website: www.wiley.com/go/sturges

The notion of using USB for control of "real-world" devices, we will assert, is based on the relatively large capacity for computes at low cost compared with specialized control hardware[3]. Through our examples, we will touch upon the necessary investments today's peripheral provider must make to remain in the business.

At the time of this writing, USB 3.0 is becoming commonplace, while we await the appearance of the next level of "wireless" communication on the mass consumer market. Compared with the legacy methods of printer port and so-called "serial port" interfaces, USB offers advantages in speed, power, and convenience[4]. Disadvantages include lack of real-time response and the need to interface with unknowable codes written by others, and purposely not publicly documented[5].

With respect to speed, USB 3.0 offers a "throughput" of 640 MBps, but the timing of such data may be uncontrollable by the user. With respect to power, a built-in 500 mA source is specified at each "port," of which there may be up to 127 such ports. Thus, some low-power peripherals may no longer require the otherwise ubiquitous "AC adapter" or wall-wart in order to function. With respect to convenience, the USB specification requires "plug and play" interconnectivity. This means, in brief, that the user may not be required to suspend the operation of other running programs while the lower-level of connectivity is negotiated by the Operating System (OS) and the peripheral. Of course, this implies that the peripheral is now required to support an intricate series of handshaking activities not required previously. Fortunately, this software task has been implemented in firmware by several vendors of "interface chips," for example, Future Technology Devices International (FTDI)[6]. On the minus side, while the USB standard is public knowledge, it spans several thousand pages of highly detailed technical data that differs in detail for each hardware implementation.[7].

## B.2 Objective

Our objective in this appendix is to enable educators, researchers, low-volume practitioners, and unique system creators to use the power of multi-GHz computing for custom programming not deemed relevant by the forces of market-driven "applications." We would note also that the need for "real-time" control still exists in laboratories. By this we refer to the practice of knowing precisely (to the microsecond) when a signal will appear at an interface for use by either party. Such real-time operations are only approximated by the alternative "synchronous service" of USB in the delivery of digital video data, but this ability is being phased out as the digital television market is supplanted by streaming technologies that do not need to support real-time transmission.

With the demise of analog television, a large market segment has also been eliminated from consideration, but real-time computing is still supported by a small cadre of laboratory-focused providers[8].

Another key objective is to educate our general engineering student population, excepting perhaps those in pursuit of degrees in Computer Science, to the means needed to achieve "software literacy" consistent with industrial automation classroom activities. We find that no market-driven "applications" address this need.

### *B.2.1 Brief Literature Review*

We acknowledge here the popularity of several alternative technologies directed to the control of real-world devices, for example, valves, relays, motors, indicators, and so on. The programmable logic controller (PLC) has been in use for decades with off-line programming methods based on then-current PCs. These devices typically employ a reduced-capacity processor for on-line implementation, performing sequential logical calculations in the same vein as parallel connected devices. In general these systems do not employ a run-time graphical user interface (GUI), and function as repetitive, customized controllers for a broad range of equipment systems with relatively low data processing tasks, for example, RS Logix[9]. With such systems, an investment needs to be made in the acquisition and practice of proprietary computer architectures and languages.

For more than 25 years, a unique software product from National Instruments[10] has enabled the creation of virtual instruments on a PC using a proprietary graphical "language." Teamed with that company's interface hardware, individual digital and analog I/O is made possible along with extensive logical internal processing. With care, one can program with procedural code embedded in the graphical media. For specialized applications, one can avoid the PC altogether and operate such virtual instruments using field-programmable gate arrays. Similar to the PLC's, an investment needs to be made in the acquisition and practice of proprietary computer architectures and languages, in addition to the I/O hardware. Recently, USB-compatible hardware for introductory use has been offered (e.g., NI's "6009" I/O block product).

We have surveyed the contemporary marketplace for introductory volumes to support the available hardware and find that a relatively high level of programming skill is assumed, while the level of hardware knowledge assumed is quite low. This indicates a trend away from inter-connection with the real-world in favor of the virtual world isolated from the exigencies of hardware save the screen and keyboard. This prospect and the market pressures that created it are well-addressed in research devoted to rapid obsolescence[11].

## B.3 Approach

Our approach to providing a kernel of knowledge which may serve the related needs of educators and the general engineering marketplace proceeds with the identification of a minimal set of interface hardware with which to establish the USB communication protocols at the "outboard" end of the information chain (Figure a2.1). We then proceed to develop the "inboard" end of the information chain with the express interest in an "open" architecture for personal program realization. In both steps, we assume the minimum amount of specialized programming and electronic knowledge of the user.

To communicate (rather than encapsulate) this knowledge, we express here the basic functions that need to be performed irrespective of the hardware or software platforms selected. Only to express examples of the method do we resort to actual coding and select a hardware platform. The functional approach was developed decades ago by teams of "value engineers" intent on reducing the cost of military acquisitions by expressing the instance of a

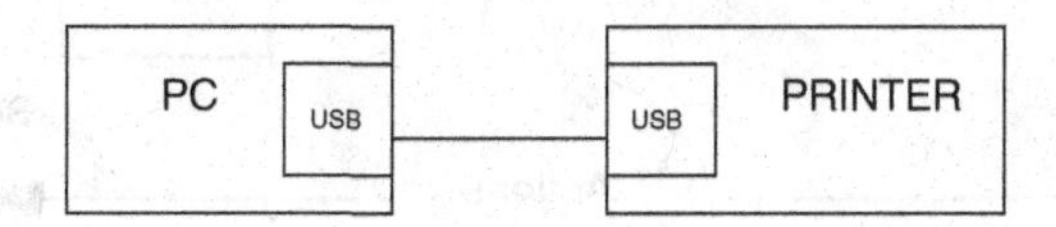


**Figure a2.1** A user's high level view of the USB information chain

product or process by its generic functions, all logically connected by "and" and "or" links[12]. An example function block diagram (FBD) is shown in Figure a2.2, wherein the highest level function is placed to the left, and its decomposed lower-level functions appear more rightward. Notice that the "flow" of the knowledge is expressed by answering the question "why" by looking towards the left, and the question "how" by looking towards the right.

The functions themselves are expressed by a single active verb and a measurable generic noun in a small rectangle. For example, "support load" is a viable function, whereas "provide support" is not[14]. There is no "right or wrong" function diagram, only viable and non-viable ones as judged by the team of users that employ them for interdisciplinary communication purposes. Please notice also that such a diagram is not intended to express the same information as a "flow chart" since the latter rarely explains the *raison d'être* for the functions and their logical relationships. In our example, the right-most series of functions, when read from the top to bottom concisely express the "pseudo code" prevalent in other articles. Taken all together, the function diagram expresses the intent of the designer or user.

In any such endeavor, one must chose software and hardware platforms to realize this intent. We have chosen the C-language for our software because it appears to be the most popular and most enduring choice. We will additionally assume a low level of expertise in C by the reader so that the essence of the project can be readily apprehended, but we will not be offering here a review of the elements of that language. We recognize the strong following of other languages such as C#, Java, and C++, but have avoided the object-oriented approach as not vital to the task at hand, and have steered away from proprietary implementations as much as possible. We have also chosen the Phidgets™ hardware platform since, for the present at least, the hardware and software

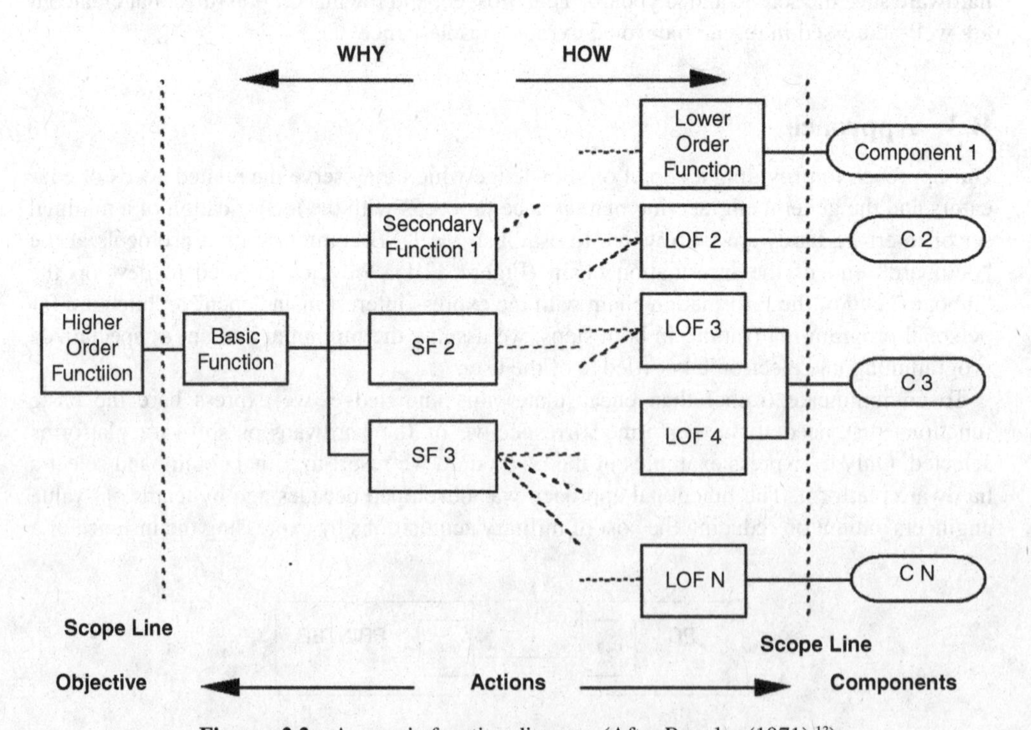


**Figure a2.2** A generic function diagram. (After Ruggles (1971).[13])

implementation is more than adequate for the inexperienced user. We are in no case affiliated with that company, and several others, such as J-Works™ or Active Wire™ may serve as well.

Another choice needed to be made in this endeavor: that of the PC itself. Since the market is led by said monopoly, we have selected Windows™ as our OS, but you will find that others support the use we make of it equally well. The PC hardware choice is fully generic, so long as it has a USB-compatible port.

To recapitulate, the goals of this project were to illustrate by practical example the steps one must take to realize communication between an application of custom-developed code as practiced by a general engineering audience with equally customized hardware. The examples we will now discuss in detail comprise a general-purpose analog/digital interface ("ifKit"), and a commonplace radio control (R/C) servo interface ("servo").

## B.4 Details of Implementation

The implementation of communication to a USB device in these cases comprises two separate instances. In the first instance, the hardware supplier has provided software that interacts with the user through a specialized window. This window displays the measured conditions of the external board at the "outside" end of the USB chain. The user cannot access the code from which this window appears, but can determine through its use the fact that the board may be attached and operating correctly. One may also "test" the outputs in some cases by setting certain fields within the window.

In the second instance, the programmer uses software calling sequences to communicate with the external board. These sequences are pre-compiled for the most part, so that again, the programmer cannot access the code through which this communication occurs.

Prior to the instantiation of either of the two instances, the user must install a set of pre-compiled codes called "drivers" into the OS of the target computer, in much the same way as "printer drivers" are installed for that purpose. Because one cannot access this code, the possibility does not exist for the user to specify the precise timing of any of the communication actions that take place. By dint of the high speed of the processor, one must assume a "level of service" that may be commensurate with the task to be performed, but there are no guarantees. Figure a2.3 illustrates the functionality of this driver (as much as can be known) by the isolated function block in the upper left-hand corner, and the software agent by the oval entitled "Phidget Software." These two fragments are connected by a solid line representing the functional "and" operator. In this example, the board is a Phidget type 1018 Interface Kit.

The instructions needed to install the drivers and to access the test window are given in the company's Programming Manual that was available on "the web" through that company's website, "Phidgets." (It is important to note that the rapidly changing nature of the hardware and software market requires the user to adopt a "digital" outlook and accept the fact that information may appear, change, and even evaporate without notice at any time. In this appendix we have avoided giving such ephemeral references wherever possible.)

Our purpose is to educate users and implement examples of the code that we do have control over, so our basic function is given as "Test Protocol." An equally valid basic function could be "Educate User" if the working team agrees. How to perform this basic function is specified by three subfunctions: Verify Communications, Display Changes, and Stop Communications. Each of these functions is further subdivided as shown.

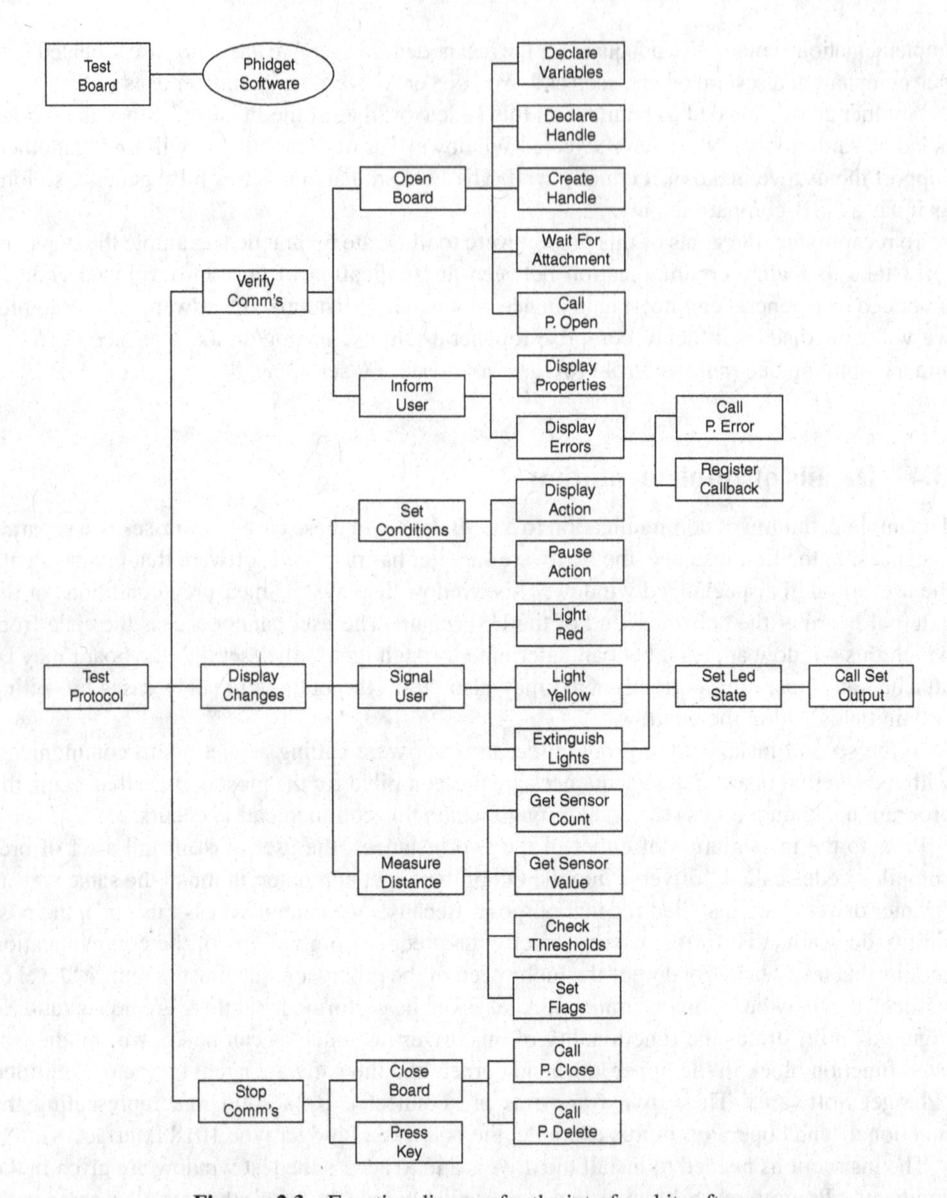


**Figure a2.3** Function diagram for the interface kit software

Taking first the Verify Communications' function, we need to "open" the board for communications, and inform the user (or his/her program) that we have done so. These two subfunctions will comprise the most novel of the tasks for the programmer with respect to the hardware. The order (top to bottom) of each subfunction for these is arbitrary, but we have arranged the FBD so that it reads at the fourth level like a flow chart vertically insofar as possible. As with any program in C, one typically declares the variables and their types at the beginning of the code. A new feature is next introduced, that of declaring a "handle" for the external device to be invoked by the code in the PC. This handle takes the place of the memory-mapped I/O of earlier days, and it is simply a pointer to an address that acts as the gateway for all communications with the device. A separate step is needed to create the space in memory for this handle. The detailed instructions for these steps are given in the on-line C/C++ examples database for this device on the Phidgets website.

To account for the possibility that the device is not physically connected to the USB port of the PC, a function is created that tests and waits for the attachment of the same for a user-specified period of time. During this time, the user may post a warning on the screen of the PC. After the negotiation for communication is made by the firmware of the device and the PC, the user may call a pre-written function to "open" the board for communication (reminiscent of the "open" command in BASIC). Each of these pre-written functions are known as the Application Programmer's Interface (API), and are supplied by the maker of the interface board so that communication with the same is facilitated. In the C language, these may be placed in so-called "header files." In this case, the major header file is named "Phidget21.h." Without it being present, the compiler will not successfully bring these functions into the user's code. Also, there will be a corresponding library of pre-compiled functions that must be linked to the compilation process. In this case the "Phidget.lib" file needs to be attached to the linker through its input dependencies listing.

As Figure a2.3 illustrates, there are functions needed to inform the user (or the user's program) of other activity: in the first subfunction, we invoke the API to fetch and display the serial number of the attached device and its physical and functional characteristics. This step is crucial if more than one of the same type of device is attached to the USB. The serial number found would be used to distinguish between the attached devices for steering the desired I/O. Each handle created would be associated with a separate serial number. In the second subfunction, error conditions are made visible, that is, an API call is made to extract details of the error condition. Also, a "call-back" function is defined to catch any errors that are thrown by the compiled code. This process involves the declaration of a function that will be invoked whenever an error condition arises. This function is of our own design but must be identified by the type "__stdcall" so that the compiler can establish a link between the internal error traps and the user's code. (There are *two* underscores before the letters "stdcall.")

The center portion of Figure a2.3 illustrates many opportunities to invoke the API provided by the Phidget company. Here, the programmer sets bits in an output register according to the numbers returned by calling analog input channels. In this case, a popular distance measurement unit (a Sharp 2Y0A02) returns an analog voltage in response to detecting a reflective object within its range. The setting of the output bits is illustrated in the C-code with concern for reader transparency of the functions, not efficiency of the code itself.

Finally, the program is shown responding to a "q" from the keyboard, indicating the end of the test. At this point, the API is again invoked with functions to close the communications and delete the memory space allocated in the handle creation step. Failure to do this may result in a build-up of inactive code, otherwise known as a "memory leak" as the host system begins to respond more slowly with repeated usage.

The entire C-code is available on our website (http://www.tbd). In its recommended form, all comments are colored green, quoted text is colored red, keywords are blue, and the code body is in black ink. Line numbers are shown in cyan. Except for the "__stdcall" and a few necessary pointers, the C-code is rendered as simply as possible for the inexperienced reader.

## *B.4.1 A Further Example*

To illustrate more fully the programming conventions needed to bridge the "USB gap" for the educator and user of real-world devices, a second example will be briefly described. In this instance a USB-driven R/C type servo board is chosen to illustrate slightly more complex communications. It is a Phidget type 1061 Advanced Servo board. Figure a2.4 shows the FBD of the interface program, and one will note how similar it is to the entirely different hardware

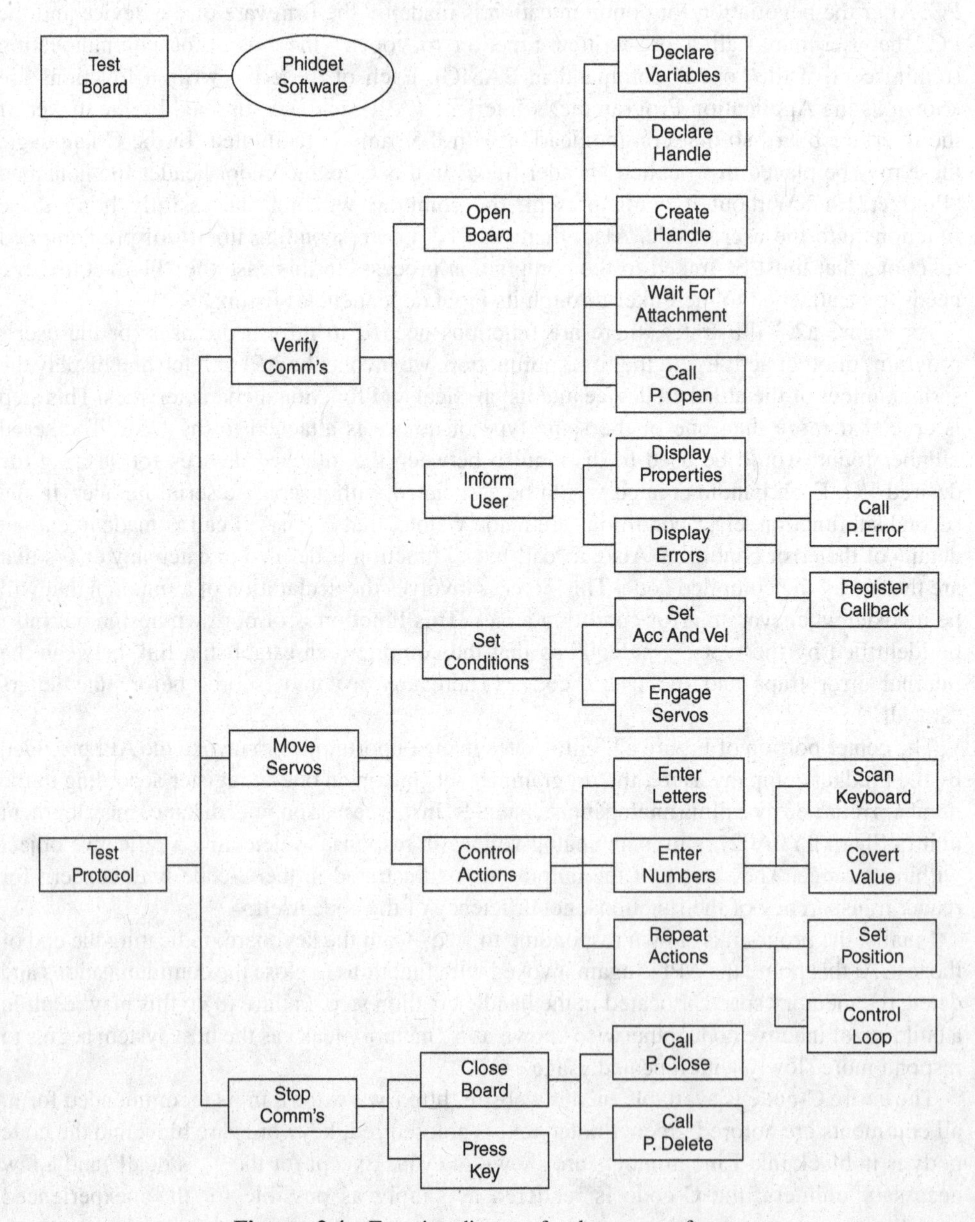


**Figure a2.4** Function diagram for the servo software

and software system represented in Figure a2.3. One of the major benefits of the FBD approach is to illustrate the commonality inherent in many artifacts that appear to be otherwise quite distinct. For this reason we will examine this example of the servo control from the viewpoint of the code itself.

The program begins (for the sake of the compiler) with the "includes." (It begins for us with the "main()" function located near the physical end.) We add <stdlib.h> since we will be using the "atof" function defined there. We continue with our global variables: a character to be read from the keyboard, a true/false indicator for the main loop, and an array of values to hold the target movements of each servo. We define eight of these, but will be using only two.

The first function is our "call-back" which is typed __stdcall. We name it ErrorHandler, and the parameter list comprises the device handle (SRV), a utility pointer that we will not be using, the ErrorCode and a pointer to the character string that describes the error itself. This function gets called by the compiled code whenever the system of Phidget API software detects a problem. Otherwise it is inactive. Like all int functions, we return a value (0) to the OS.

Our next function, display_properties, needs one parameter of type "CPhidgetAdvancedServoHandle." The C language encourages customized typing, and in this case makes it clear that the language is C, the vendor is Phidget, the type of interface board is AdvancedServo, and that the parameter itself is the pointer/handle to the gateway that transmits all information about the board to the program which uses it. Our local variables include facts about the board (to be determined) and a character string pointer for storing the name of the board. We then make a call to the API function to return the type of device connected. Its parameter list contains a seldom-used feature: the handle is "cast" as a "CPhidgetHandle" type just to be sure that the compiler knows to look in its … phidget21 library. A list of this information serves as the initial data base for identifying the device, should there be multiple copies of the same type of board connected to the USB.

Our next function, move_servo, needs the familiar handle, SRV. This small function reads the keyboard to get a floating-point number by way of a character string of up to 10 characters (value). After scanning the keyboard, this string is converted to a double and saved in an array for use in the next line. We then call another API function which needs the handle, the index value of the servo motor to move, and the exact angle in degrees we command it to move to. Incidentally, the list of all the API functions available does not appear in any reference book, but rather "on-line" at the Phidget website. The interested reader must navigate to this list and its recommended usages by selecting the specific Phidget device from a list (in the left-hand column). The list is only a few steps below the home page, so we will not describe it in detail here.

The "big" function appears next, and it would be considered the "main" function but for the stylistic approach to C which keeps main mostly empty. Line 49 is both a declaration and an assignment, making space for the device handle. In older PCs this would be a fixed location in memory to which we would make all references to our attached device. The next line invokes an API function to actually reserve space for all the data associated with the handle and our attached device. We then call an API function to define the name of the error handler we defined earlier with our __stdcall. Now, the compiled code will compute the correct reference to invoke when an error is thrown. Finally, we call another API function to open the board for communications. The "–1" parameter is a requirement of the API.

We now ask if the device is indeed connected via USB (has the interface been negotiated satisfactorily?), and wait for a prescribed number of milliseconds if not. If no problems occurred, we collect and display the properties of the connected device. A for-loop follows which sets the properties for each servo dynamic response, and chose at this point to actually engage the servo into action. This last step can be inserted anywhere appropriate for our needs.

A user-terminated while-loop appears next, prompting the user to enter an "m" to move a servo, or a "q" to quit. The switch statement uses the ASCII equivalent of these characters. If we do elect to quit, we invoke a pair of API calls to close the board from communication and delete the memory allocated for it. The "main" that follows is the actual entry point for a C program.

## B.5 Results

The practical result of this project is the implementation of a simple, low-cost demonstration board for the educator and first-time user of customized USB devices. Figure a2.5 depicts a 5×8 in. platform upon which has been attached the two USB interface boards mentioned above. Also included are a pair of infra-red (IR) distance measuring units and a pair of R/C servo units. The InterfaceKit 1018 also has a pair of LEDs attached to two of its eight digital output ports. Not shown are the USB cables that attach to the host PC and the small power supply needed to power the 1061 board servos.

The intellectual result of this project is the understanding of the user-designed and written software which works through the "encapsulated" codes of the maker of the boards and the authors of an OS and C compiler for a PC. We have shown how to establish communication with USB-powered interface boards of differing functions with a few fundamental programming ideas. We also illustrate the use of a function-defined structure for designing the code at a high level before committing to writing at the level of the compiler. In this way, the sections can be examined by all stakeholders and agreement reached before the debugging process begins.

**Figure a2.5** Instructional board for the interface kit and the servo

## B.6 Conclusion

This appendix has endeavored to provide implementation transparency for the developer and user of low-quantity, often unique interfaces without the need for specialists in "information technology" and computer networking. We look forward to the application of this work to industrial automation classroom activities, at least, and to implementations within the research communities far removed from "computer science." We have found that one hardware and software system worked very well, and could work across many platforms, but caution that many others exist and that we make no recommendations or endorsements of a particular vendor's product.

We also recommend carefully exploring the API provided by the maker before choosing a USB interface board for your own purposes. Your choice of language may not be C, but the API you select must be compilable from your language of choice.

c:\Documents and Settings\sturges\Desktop...InterfaceKit-dense\InterfaceKit-dense.cpp 1

```
1 // - InterfaceKit dense - This version reads the sensors and lights the LEDS
     accordingly
2
3 #include "stdafx.h"          //a need of Visual Studio
4 #include "phidget21.h"       //all the API functions are in here
5
6 int __stdcall ErrorHandler(CPhidgetHandle IFK, void *userptr, int ErrorCode, const char
     *unknown)
7 {   //This is an event-triggered call-back function
8     if(ErrorCode == 14) return 0;
9     printf("Error handled. %d - %s", ErrorCode, unknown);
10    return 0;   //note that "ErrorHandler" is defined in line 66
11 }
12 //Display the properties of the attached phidget to the screen.
13 //We will be displaying the name, serial number and version of the attached device.
14 //Will also display the number of inputs, outputs, and analog inputs on the interface
     kit
15 int display_properties(CPhidgetInterfaceKitHandle phid) //tell all
16 {
17    int serialNo, version, numInputs, numOutputs, numSensors;
18    const char* ptr;        //need a char string to display type of Phidget
19    CPhidget_getDeviceType((CPhidgetHandle)phid, &ptr);
20    printf("%s\n", ptr);    //ALL char strings are referenced with a pointer
21    CPhidget_getSerialNumber((CPhidgetHandle)phid, &serialNo);
22    CPhidget_getDeviceVersion((CPhidgetHandle)phid, &version);
23    printf("Serial Number: %10d\nVersion: %8d\n", serialNo, version);
24    CPhidgetInterfaceKit_getInputCount(phid, &numInputs);
25    CPhidgetInterfaceKit_getOutputCount(phid, &numOutputs);
26    printf("# Digital Inputs: %d\n# Digital Outputs: %d\n", numInputs, numOutputs);
27    CPhidgetInterfaceKit_getSensorCount(phid, &numSensors);
28    printf("# Sensors: %d\n", numSensors);//but we'll only use 2
29    printf("Press 'q' to quit. ");  //a way to exit gracefully
30    return 0;
31 }
32 int SetLEDState(CPhidgetInterfaceKitHandle ifKit, int number, int state)
33 {           //setup for poking bits into the output byte
34    CPhidgetInterfaceKit_setOutputState(ifKit, number, state);
35    return 0;
36 }
37 int RedOn(CPhidgetInterfaceKitHandle ifKit)
38 {           //poke bit into the output byte
39        SetLEDState(ifKit, 1, 1);
40        return 0;
41 }
42 int YellowOn(CPhidgetInterfaceKitHandle ifKit)
43 {           //poke bit into the output byte
44        SetLEDState(ifKit, 0, 1);
45        return 0;
46 }
```

```
47 int AllOff(CPhidgetInterfaceKitHandle ifKit)
48 {        //poke bits into theoutput bytes
49       SetLEDState(ifKit, 0, 0);
50       SetLEDState(ifKit, 1, 0);
51       return 0;
52 }
53 int GetSensorValue(CPhidgetInterfaceKitHandle ifKit, int channel)
54 {        //set up to read a sensor
55    int value =0;
56    CPhidgetInterfaceKit_getSensorValue(ifKit, channel, &value);//get it
57    return value;
58 }
59 int interfacekit_working()
60 {  //all the action happens in this function
61    int result, numSensors, i, value, Yellowflag, Redflag;
62    int val[8] = {0};    //up to 8 sensors
63    const char *err;     //just in case an error happens
```

c:\Documents and Settings\sturges\Desktop...InterfaceKit-dense\InterfaceKit-dense.cpp 2

```
64     char get_char = 'a'; //random starter value
65     printf("This version reads the sensors and lights the LEDS accordingly.\n");
66     CPhidgetInterfaceKitHandle ifKit = 0;    //Declare an InterfaceKit handle
67     CPhidgetInterfaceKit_create(&ifKit);     //create the InterfaceKit object
68     CPhidget_set_OnError_Handler((CPhidgetHandle)ifKit, ErrorHandler, NULL);//define
       the handler
69     CPhidget_open((CPhidgetHandle)ifKit, -1);//open the interfacekit for device
       connections
70     printf("Waiting for interface kit to be attached...\n");
71     if((result = CPhidget_waitForAttachment((CPhidgetHandle)ifKit, 10000)))//10 seconds
72     {         //give the user 10 seconds to connect the I/O device
73        CPhidget_getErrorDescription(result, &err); //get error type
74        printf("Problem waiting for attachment: %s\n", err);
75        return 0;
76     }
77     display_properties(ifKit);  //strut your stuff
78     printf("Ready...\n");   //indicate readiness
79     while(get_char != 'q')  //do while not ready to quit
80 {
81     printf("Press CR to read sensors. \n");
82     get_char = getchar();    //if CR, continue; if 'q' quit next time
83     printf("Reading sensors....\n");
84     CPhidgetInterfaceKit_getSensorCount(ifKit, &numSensors);    //get the number of
       sensors
85     for(i = 0; i < 2; i++)  //let's just use 2 for now
86     {
87        val[i] = GetSensorValue(ifKit, i);  //Read a sensor into array
88        printf("Sensor %d reads %d\n",i,val[i]);//show result
89     }
90     Yellowflag = 0; Redflag = 0;             //clear flags
91     for(i=0; i<2; i++)  //we'll just use 2 sensors for now
92     {
93       if(val[i]>=400) Yellowflag=1;          //check distance
94       if(val[i]>=500) Redflag=1;             //check closer distance
95     }
96     AllOff(ifKit);                           //lights out = reset
97     if(Redflag) RedOn(ifKit);                //red light
98     else if(Yellowflag) YellowOn(ifKit);     //yellow light
99 }   //end of while loop; keep doing this forever
100    //keep displaying interface kit data until 'q' is read
101    printf("Closing...\n");
102    CPhidget_close((CPhidgetHandle)ifKit);//so we will close the phidget
103    CPhidget_delete((CPhidgetHandle)ifKit);//and delete the object we created
104    return 0;          //all done, exit
105 }
106 int main()
107 {
108    interfacekit_working(); //just a style to keep main free
109    return 0;
110 }
111
```

## Notes

1. Bergsman, P. (1994,) *Controlling the World with Your PC*, HighText Publications.
2. Negroponte, N. (1995) *Being Digital*, Knopf.
3. Gleik, J. (2000) *Faster*, Vintage Publishing.
4. Podgórski, A., Nedwidek, R. and Pochmara, M. (2003) Implementation of the USB Interface in the Instrumentation for Sound and Vibration Measurements and Analysis. IEEE International Workshop on Intelligent Data Acquisition and Advanced Computing Systems: Technology and Applications, September 8–10, 2003, Lviv, Ukraine, pp. 159–163.
5. Horton, I. (2008) *Beginning Visual C++ 2008*.
6. www.ftdichip.com (accessed November 3, 2011).
7. *Universal Serial Bus Specification*, Revision 1.1. www.usb.org/home (accessed June 18, 2014).
8. www.ni.com/labview/ (accessed November 3, 2011).
9. www.rockwellautomation.com/rockwellsoftware/design/rslogix5/ (accessed November 3, 2011).
10. Bishop, R. (2001) *Learning with LabVIEW 6i*, Prentice Hall.
11. Zheng, L. (2010) Ontology-based knowledge representation and decision support for managing product obsolescence. Ph.D. dissertation. Virginia Polytechnic Institute and State University.
12. Fowler, T. C. (1990) *Value Analysis in Design*, Van Nostrand Reinhold.
13. Ruggles, W.F. (1971) FAST—a management planning tool. *SAVE Encyclopedia of Value*, **6**,301.
14. Sturges, R.H., O'Shaughnessy, K. and Reed, R.G. (1993) A systematic approach to conceptual design based on function logic. *International Journal of Concurrent Engineering: Research and Applications*, **1** (2), 93–106.

# Appendix C

# Microchip Code for USB Board to PPM Translation

```
                ;LMM3.SRC         14 SEP 12
                ;This code translates PWM from the Phidget 1061
                ;into PPM that the transmitter needs
                ;adds a sync output to channel 4 every 4th time
        DEVICE  PIC12C509,IRC_OSC,WDT_OFF,PROTECT_OFF,MCLR_OFF
        RESET   SETIO
OUT     EQU     GP0          ;Bitstring out
IN      EQU     GP1          ;Bitstring in
SYNC    EQU     GP2          ;Sync out
W1      EQU     11H          ;1st width in 10's of usec
W2      EQU     12H          ;2nd width in 10's of usec
W3      EQU     13H          ;3rd width in 10's of usec
W4      EQU     14H          ;4th width in 10's of usec
T       EQU     15H          ;timer byte
TIME    EQU     16H          ;saved time value

SETIO   CLR     GPIO         ;clean slate
        MOV     !OPTION,     #11011111B     ;clear T0CS
        MOV     !GPIO,       #11111010B     ;two output bits

START   SETB    OUT          ;say hi by setting out to ready state
WFI1    JNB     IN,WFI1      ;idle while the input line is low
        CLR     W1           ;the 1st width time
INC1    INC     W1           ;start the count
        CALL    WAIT         ;wait 8 usec here
        JB      IN,INC1      ;if still high, continue counting
```

*Practical Field Robotics: A Systems Approach*, First Edition. Robert H. Sturges, Jr.

Companion Website: www.wiley.com/go/sturges

```
WFI2   JNB   IN,WFI2   ;idle while the input line is low
       CLR   W2        ;the 2nd width time
INC2   INC   W2        ;start the count
       CALL  WAIT      ;wait 8 usec here
       JB    IN,INC2   ;if still high, continue counting
WFI3   JNB   IN,WFI3   ;idle while the input line is low
       CLR   W3        ;the 3rd width time
INC3   INC   W3        ;start the count
       CALL  WAIT      ;wait 8 usec here
       JB    IN,INC3   ;if still high, continue counting
WFI4   JNB   IN,WFI4   ;idle while the input line is low
       CLR   W4        ;the 4th width time
INC4   INC   W4        ;start the count
       CALL  WAIT      ;wait 8 usec here
       JB    IN,INC4   ;if still high, continue counting

       CLRB  OUT       ;start sending the string out
       CALL  SPACE     ;start bit time, about 300 usec
       SETB  OUT       ;start width #1
       MOV   TIME,W1   ;save width
       CALL  PULSE     ;time it like WAIT
       CLRB  OUT       ;start sending the next bit out
       CALL  SPACE     ;start bit time, about 300 usec
       SETB  OUT       ;start width #2
       MOV   TIME,W2   ;save width
       CALL  PULSE     ;time it like WAIT
       CLRB  OUT       ;start sending the next bit out
       CALL  SPACE     ;start bit time, about 300 usec
       SETB  OUT       ;start width #3
       MOV   TIME,W3   ;save width
       CALL  PULSE     ;time it like WAIT
       CLRB  OUT       ;start sending the last bit out
       CALL  SPACE     ;start bit time, about 300 usec

       SETB  SYNC      ;send a sync pulse
       CALL  SPACE     ;bit time, about 300 usec
       CLRB  SYNC      ;clear the sync pulse

       SETB  OUT       ;start width #4
       MOV   TIME,#200 ;save width
       CALL  PULSE     ;time it like WAIT
       CLRB  OUT       ;send the stop bit out
       CALL  SPACE     ;start bit time, about 300 usec
       SETB  OUT       ;all done!

WFI11  JNB   IN,WFI11  ;idle while the input line is low
       CLR   W1        ;the 1st width time
```

```
INC11   INC     W1          ;start the count
        CALL    WAIT        ;wait 8 usec here
        JB      IN,INC11    ;if still high, continue counting
WFI12   JNB     IN,WFI12    ;idle while the input line is low
        CLR     W2          ;the 2nd width time
INC12   INC     W2          ;start the count
        CALL    WAIT        ;wait 8 usec here
        JB      IN,INC12    ;if still high, continue counting
WFI13   JNB     IN,WFI13    ;idle while the input line is low
        CLR     W3          ;the 3rd width time
INC13   INC     W3          ;start the count
        CALL    WAIT        ;wait 8 usec here
        JB      IN,INC13    ;if still high, continue counting
WFI14   JNB     IN,WFI14    ;idle while the input line is low
        CLR     W4          ;the 4th width time
INC14   INC     W4          ;start the count
        CALL    WAIT        ;wait 8 usec here
        JB      IN,INC14    ;if still high, continue counting

        CLRB    OUT         ;start sending the string out
        CALL    SPACE       ;start bit time, about 300 usec
        SETB    OUT         ;start width #1
        MOV     TIME,W1     ;save width
        CALL    PULSE       ;time it like WAIT
        CLRB    OUT         ;start sending the next bit out
        CALL    SPACE       ;start bit time, about 300 usec
        SETB    OUT         ;start width #2
        MOV     TIME,W2     ;save width
        CALL    PULSE       ;time it like WAIT
        CLRB    OUT         ;start sending the next bit out
        CALL    SPACE       ;start bit time, about 300 usec
        SETB    OUT         ;start width #3
        MOV     TIME,W3     ;save width
        CALL    PULSE       ;time it like WAIT
        CLRB    OUT         ;start sending the last bit out
        CALL    SPACE       ;start bit time, about 300 usec
        SETB    OUT         ;start width #4
        MOV     TIME,#30    ;save width
        CALL    PULSE       ;time it like WAIT
        CLRB    OUT         ;send the stop bit out
        CALL    SPACE       ;start bit time, about 300 usec
        SETB    OUT         ;all done!

WFI21   JNB     IN,WFI21    ;idle while the input line is low
        CLR     W1          ;the 1st width time
INC21   INC     W1          ;start the count
        CALL    WAIT        ;wait 8 usec here
```

```
        JB      IN,INC21  ;if still high, continue counting
WFI22   JNB     IN,WFI22  ;idle while the input line is low
        CLR     W2        ;the 2nd width time
INC22   INC     W2        ;start the count
        CALL    WAIT      ;wait 8 usec here
        JB      IN,INC22  ;if still high, continue counting
WFI23   JNB     IN,WFI23  ;idle while the input line is low
        CLR     W3        ;the 3rd width time
INC23   INC     W3        ;start the count
        CALL    WAIT      ;wait 8 usec here
        JB      IN,INC23  ;if still high, continue counting
WFI24   JNB     IN,WFI24  ;idle while the input line is low
        CLR     W4        ;the 4th width time
INC24   INC     W4        ;start the count
        CALL    WAIT      ;wait 8 usec here
        JB      IN,INC24  ;if still high, continue counting

        CLRB    OUT       ;start sending the string out
        CALL    SPACE     ;start bit time, about 300 usec
        SETB    OUT       ;start width #1
        MOV     TIME,W1   ;save width
        CALL    PULSE     ;time it like WAIT
        CLRB    OUT       ;start sending the next bit out
        CALL    SPACE     ;start bit time, about 300 usec
        SETB    OUT       ;start width #2
        MOV     TIME,W2   ;save width
        CALL    PULSE     ;time it like WAIT
        CLRB    OUT       ;start sending the next bit out
        CALL    SPACE     ;start bit time, about 300 usec
        SETB    OUT       ;start width #3
        MOV     TIME,W3   ;save width
        CALL    PULSE     ;time it like WAIT
        CLRB    OUT       ;start sending the last bit out
        CALL    SPACE     ;start bit time, about 300 usec
        SETB    OUT       ;start width #4
        MOV     TIME,#30  ;save width
        CALL    PULSE     ;time it like WAIT
        CLRB    OUT       ;send the stop bit out
        CALL    SPACE     ;start bit time, about 300 usec
        SETB    OUT       ;all done!

WFI31   JNB     IN,WFI31  ;idle while the input line is low
        CLR     W1        ;the 1st width time
INC31   INC     W1        ;start the count
        CALL    WAIT      ;wait 8 usec here
        JB      IN,INC31  ;if still high, continue counting
```

```
WFI32   JNB     IN,WFI32   ;idle while the input line is low
        CLR     W2         ;the 2nd width time
INC32   INC     W2         ;start the count
        CALL    WAIT       ;wait 8 usec here
        JB      N,INC32    ;if still high, continue counting
WFI33   JNB     IN,WFI33   ;idle while the input line is low
        CLR     W3         ;the 3rd width time
INC33   INC     W3         ;start the count
        CALL    WAIT       ;wait 8 usec here
        JB      IN,INC33   ;if still high, continue counting
WFI34   JNB     IN,WFI34   ;idle while the input line is low
        CLR     W4         ;the 4th width time
INC34   INC     W4         ;start the count
        CALL    WAIT       ;wait 8 usec here
        JB      IN,INC34   ;if still high, continue counting

        CLRB    OUT        ;start sending the string out
        CALL    SPACE      ;start bit time, about 300 usec
        SETB    OUT        ;start width #1
        MOV     TIME,W1    ;save width
        CALL    PULSE      ;time it like WAIT
        CLRB    OUT        ;start sending the next bit out
        CALL    SPACE      ;start bit time, about 300 usec
        SETB    OUT        ;start width #2
        MOV     TIME,W2    ;save width
        CALL    PULSE      ;time it like WAIT
        CLRB    OUT        ;start sending the next bit out
        CALL    SPACE      ;start bit time, about 300 usec
        SETB    OUT        ;start width #3
        MOV     TIME,W3    ;save width
        CALL    PULSE      ;time it like WAIT
        CLRB    OUT        ;start sending the last bit out
        CALL    SPACE      ;start bit time, about 300 usec
        SETB    OUT        ;start width #4
        MOV     TIME,#30   ;save width
        CALL    PULSE      ;time it like WAIT
        CLRB    OUT        ;send the stop bit out
        CALL    SPACE      ;start bit time, about 300 usec
        SETB    OUT        ;all done!
        JMP     START      ;do it again

WAIT    NOP     ;use 6 no-ops for 10 usec total
WAIT1   NOP     ;just in case we need only 5
        NOP
        NOP
        NOP
```

```
       NOP
       RET

SPACE  MOV     T,#100    ;guess at 300 usec total
SPACE1 DECSZ   T         ;count down to zero
       JMP     SPACE1    ;not done yet
       RET

PULSE  MOV     T,TIME    ;the width of the pulse measured
PULSE1 CALL    WAIT      ;entry point for decrement
       DECSZ   T         ;count down to zero
       JMP     PULSE1    ;not done yet
       RET
```

# Appendix D

## Selected Electronic Parts for Mowing Robot

We suggest that you find the following data sheets on the Internet to be sure of the electrical specifications and physical packages.

| LM695 | Power Op Amp |
|---|---|
| 250ST180/250SR180 | Air Ultrasonic Ceramic Transducers |
| Phidgets™1047 board | Encoder High Speed 4-input |
| Phidgets™1061 board | Advanced Servo 8-motor |
| Sabertooth™ 2X25V2 | Dual 25 Amp Motor Driver |
| LM358 | Dual Low Power Op Amp |
| LM317 | Variable voltage regulator, positive |
| CD4001 | CMOS Quad NOR gate |
| CD4013 | CMOS Dual D-Flipflop |
| CD40106 | CMOS Hex Inverter, Schmitt Trigger |
| 12C509 | MicroChip™ Programmable Integrated Circuit (programmable with MPLAB™ software and any of many chip burners) |
| LM567 | IC Tone Decoder |
| LM883 | Dual High Speed Audio Op Amp |
| LT1115 | Ultra-low Noise Audio Op Amp |

*Practical Field Robotics: A Systems Approach*, First Edition. Robert H. Sturges, Jr.

Companion Website: www.wiley.com/go/sturges

# Appendix E

## Software Concordance

| Function | Calls | Called by |
|---|---|---|
| screen_init() | initscr()<br>cbreak()<br>refresh()<br>wrefresh()<br>newwin()<br>box() | main()<br>stop_servo()<br>move_servo(). |
| initscr() | curses.h | screen_init() |
| cbreak() | curses.h | screen_init() |
| refresh() | curses.h | screen_init() |
| wrefresh() | curses.h | screen_init() |
| newwin() | curses.h | screen_init() |
| box() | curses.h | screen_init() |
| curs_set() | curses.h | update_display2()<br>update_display3()<br>update_display4()<br>update_display_servo() |
| mvwprint() | curses.h | update_display2()<br>update_display3()<br>update_display4()<br>update_display_servo() |
| wrefresh() | curses.h | update_display2()<br>update_display3()<br>update_display4()<br>update_display_servo() |
| endwin() | curses.h | screen_end() |

(*Continued*)

*Practical Field Robotics: A Systems Approach*, First Edition. Robert H. Sturges, Jr.

Companion Website: www.wiley.com/go/sturges

| Function | Calls | Called by |
|---|---|---|
| update_display2() | curs_set()<br>mvwprint()<br>wrefresh() | get_median()<br>InputHandlerE() |
| update_display3() | curs_set()<br>mvwprint()<br>wrefresh() | pushdown()<br>lost()<br>waypoint_drive() |
| update_display4() | curs_set()<br>mvwprint()<br>wrefresh() | push4()<br>lost()<br>waypoint_drive() |
| update_display_<br>servo() | curs_set() | stop_servo() |
| | mvwprint()<br>wrefresh() | move_servo()<br>servo_go() |
| screen_end() | endwin() | main() |
| InputHandlerE() | CPhidgetEncoder_<br>getIndexPosition()<br>Beep()<br>find_line()<br>push()<br>pushflag()<br>update_display2() | *interrupt* |
| Beep() | Windows.h | test_servo()<br>locate()<br>InputHandlerE()<br>lost()<br>waypoint_drive() |
| CPhidgetEncoder_<br>getIndexPosition() | phidget21.h | InputHandlerE() |
| ErrorHandlerS() | | *interrupt* |
| ErrorHandlerE() | | *interrupt* |
| dist2line() | | waypoint_drive()<br>locate() |
| push() | | InputHandlerE() |
| pushflag() | | InputHandlerE() |
| push4() | update_display4() | locate() |
| pushdown() | update_display3() | locate()<br>waypoint_drive() |
| principal() | | waypoint_drive()<br>get_ze() |
| get_line() | | get_jstar() |
| get_perp() | | waypoint_drive() |
| get_<br>intersection() | | get_jstar() |
| get_median() | screen_init()<br>update_display2()<br>Sleep() | locate()<br>main() |
| stop_servo() | CPhidgetAdvancedServo_<br>setPosition()<br>screen_init() | move_servo()<br><br>locate() |

| Function | Calls | Called by |
|---|---|---|
| | update_display_servo() | startout()<br>waypoint_drive()<br>lost()<br>main() |
| servo_go() | CPhidgetAdvancedServo_setPosition()<br>update_display_servo() | waypoint_drive()<br>main()<br>locate()<br>startout() |
| test_servo() | Beep()<br>CPhidgetAdvancedServo_setPosition() | main() |
| move_servo() | stop_servo()<br>CPhidgetAdvancedServo_setPosition()<br>screen_init()<br>update_display_servo() | main() |
| find_line() | | InputHandlerE() |
| find_point() | stnd_dev() | |
| find_dir() | | get_ze() |
| find_diff() | | |
| find_ze() | | |
| get_jstar() | get_line()<br>get_intersection() | waypoint_drive() |
| extrapolate() | | locate() |
| triangulate() | | locate() |
| get_ze() | find_dir()<br>get_heading() | |
| smoother() | | locate() |
| locate() | stop_servo()<br>get_median()<br>triangulate()<br>extrapolate()<br>Beep()<br>pushdown()<br>smoother()<br>push4() | waypoint_drive()<br>main()<br>startout()<br>lost()<br>time_reads() |
| get_heading() | | get_ze()<br>waypoint_drive() |
| startout() | Sleep()<br>stop_servo() | waypoint_drive() |
| Sleep() | Windows.h | startout()<br>get_median()<br>waypoint_drive()<br>locate() |
| stnd_dev() | | find_point()<br>xyzwrite() |
| filewrite() | | waypoint_drive() |
| fileread() | | waypoint_drive() |

(*Continued*)

| Function | Calls | Called by |
|---|---|---|
| xyzwrite() | stnd_dev() | //waypoint_drive() |
| lost() | stop_servo()<br>locate()<br>update_display4()<br>update_display3() | //waypoint_drive() |
| waypoint_drive() | stop_servo()<br>fileread()<br>Sleep()<br>get_heading()<br>get_jstar()<br>get_perp()<br>dist2line()<br>clear()<br>CPhidgetAdvancedServo_setAcceleration()<br>servo_go()<br>locate()<br>filewrite()<br>Beep() | main() |
| clear() | curses.h | get_median()<br>waypoint_drive() |
| CPhidgetAdvancedServo_setAcceleration() | phidgets21.h | waypoint_drive()<br>open_servos() |
| receivers_out() | | main() |
| receivers_in() | | main() |
| display() | | |
| time_reads() | GetSystemTime() | main() |
| GetSystemTime() | Windows.h | time_reads() |
| open_encoders() | CPhidgetEncoder_create()<br>CPhidget_set_OnErrorHandler()<br>CPhidgetEncoder_set_OnInputChangeHandler()<br>CPhidget_open()<br>CPhidget_waitForAttachment()<br>CPhidget_getErrorDescription()<br>CPhidgetEncoder_setEnabled() | main() |
| CPhidgetEncoder_create() | phidgets21.h | open_encoders() |
| CPhidget_set_OnErrorHandler() | phidgets21.h | open_encoders()<br>open_servos() |

| Function | Calls | Called by |
|---|---|---|
| CPhidget Encoder_set_OnIn-put ChangeHandler() | phidgets21.h | open_encoders() |
| CPhidget_open() | phidgets21.h | open_encoders() open_servos() |
| CPhidget_wait ForAttachment() | phidgets21.h | open_encoders() open_servos() |
| CPhidget_get Error Description() | phidgets21.h | open_encoders() open_servos() |
| CPhidgetEncoder_ setEnabled() | phidgets21.h | open_encoders() |
| display_ Eproprties() | CPhidget_getDevice-Type() CPhidget_getSerial Number() CPhidget_getDevice Version() CPhidgetEncoder_get EncoderCount() | main() |
| CPhidget_ getDeviceType() | phidgets21.h | display_Eproprties() |
| CPhidget_get SerialNumber() | phidgets21.h | display_Eproprties() |
| CPhidget_get DeviceVersion() | phidgets21.h | display_Eproprties() |
| CPhidgetEncoder_ getEncoderCount() | phidgets21.h | display_Eproprties() |
| open_servos() | CPhidgetAdvancedServo_ create() CPhidget_set_OnError Handler() CPhidget_open() CPhidget_waitFor Attachment() CPhidget_getError Description() CPhidgetAdvancedServo_ setAcceleration() CPhidgetAdvancedServo_ setVelocityLimit() CPhidgetAdvancedServo_ setSpeedRampingOn() CPhidgetAdvancedServo_ setEngaged() | main() |

(*Continued*)

| Function | Calls | Called by |
|---|---|---|
| | phidgets21.h | |
| CPhidget AdvancedServo_ create() | phidgets21.h | open_servos() |
| CPhidget AdvancedServo_ setAcceleration() | phidgets21.h | open_servos() |
| CPhidget AdvancedServo_ setVelocity Limit() | phidgets21.h | open_servos() |
| CPhidget AdvancedServo_set SpeedRampingOn() | phidgets21.h | open_servos() |
| CPhidget AdvancedServo_ setEngaged() | phidgets21.h | open_servos() |
| display_ Sproprties() | CPhidget_getDevice-Type()<br>CPhidget_getSerial Number()<br>CPhidget_getDevice Version()<br>CPhidgetAdvancedServo_ getMotorCount() | main() |
| CPhidget AdvancedServo_ getMotorCount() | phidgets21.h | display_Sproprties() |
| main() | screen_init()<br>open_encoders()<br>display_Eproperties()<br>open_servos()<br>display_Sproperties()<br>stop_servo()<br>update_display3()<br>receivers_out()<br>get_median()<br>servo_go()<br>locate()<br>move_servo()<br>fileread()<br>GetSystemTime()<br>waypoint_drive()<br>test_servo()<br>CPhidget_close()<br>CPhidget_delete()<br>screen_end() | *system* |

# Appendix F

# Solutions

## Chapter 1

1.1 Like the sewing machine, the USPS has considered pick-and-place robots in imitation of manual processes, but generally uses systems of conveyors, mechanical agitators, and sizing gates to separate the mail streams. Once letters and flats are separated, optical systems read the addresses and sort again into readable and non-readable streams. Bulk mailers employ carefully stacked cartons of mail in standard sizes so that machine handling becomes very efficient. The rise in use of electronic mail has made some of these processes obsolete, but has transferred some of the work of these transactions to the consumer.

1.2 Any of the many move and grasp amusement devices in which the user controls a small gantry crane to pick up prizes and deliver them to an output chute. These typically employ a smooth-surfaced "gripper." Another possible answer is the Space Shuttle's "Canadarm." This device must deal with high inertial loads and inherent flexibility of the members.

1.3 The first (and many present) field robots of this type employ a laser distance measurement device (LIDAR) and at least one video capture and analysis system. Current self-parking automobiles employ SODAR devices, scanning their environments for reflected sound waves to determine "free and open" areas for maneuvering. Also, computer-based vision algorithms are being used to detect the distance between moving vehicles for automatic separation control. Such systems must operate under "noisy" environmental conditions.

1.4 A valid FBD may occur in many forms, but should never contain a linear sequence of events, as with a flow chart. An FBD must comprise branching of functions connected by "and" links and rectangular function "blocks" containing a single *active* verb and a single *measurable* noun.

*Practical Field Robotics: A Systems Approach*, First Edition. Robert H. Sturges, Jr.

Companion Website: www.wiley.com/go/sturges

## Chapter 2

2.1 Environment: Figure 2.1, Figure 2.2, Figure 2.3, Figure 2.4, Figure 2.11; work: Figure 2.7, Figure 2.10, Figure 2.14.

2.2 Halogens are the chemical elements fluorine, chlorine, bromine, iodine, and astatine. Any form of these chemicals are suspect for reaction with the metals of the plant. They are also known as "salt generators" because they vigorously react with metals.

2.3 Each pin of the walker is designed with an extra "lift" mechanism so that each pin jams tightly to the tubesheet every time.

2.4 Each pin actuation mechanism has a lateral "float" or lost-motion that can be forcefully re-centered after pin insertion. An alternative could be the addition of a Remote Center Compliance device.

2.5 The arm needs to raise the walker plus its own weight. Assuming that each of three arm segments weighs 16 kg and finding the center of mass to be at about 750 mm, we calculate a torque of about 600 Nm.

2.6 The higher level function is Control Positions and this FBD should begin with Feedback positions. It should be followed with a functional map of the devices in Figure 2.16. This functional map should comprise, at least, the following: Display Vertical/Horizontal/ Distance, Detect Vertical/Horizontal/ Distance, Drive Display, Detect Shadow/ Silhouette. Other functions may be included, but the rubrics of the "and" links, "how" reading from left to right, and "why" reading from right to left must be present. Please see Appendix A for details on FBD construction.

2.7 Angles do not scale, only lengths do. All distance relationships can be considered to be similar triangles.

## Chapter 3

3.1 If the height of the coal seam varied significantly or was less than about 2 m, the room-and-pillar approach would have greater flexibility.

3.2 If the height of the coal seam was less than about 1.5 m or pitched significantly from the horizontal, independently driven coal scoops would have greater flexibility.

3.3 At minimum, the angle between each of two bridge conveyors (item 30 in Figure 3.7) and the supporting MBC, and the slack distance need to be continuously measured. Also, the distance to and the azimuth angle relative to the mine walls and the MBC need to be monitored at all times. In addition, many sensors for detecting "normal" operation need to signal to the operator of the continuous miner at the head of this kinematic chain.

3.4 A design goal is to minimize the number of MBCs and therefore maximize the length of each bridge conveyor (item 30). Long pin-to-pin distances constrain the paths in confined spaces, that is, between existing mine walls.

3.5 At minimum, any tracked vehicle must slip laterally in order to turn; the mine floor is generally covered with ground rock (sand or gravel) and coal itself; in some locations ground water forms pools that reduce the friction between the steel tracks of the MBC and the mine floor; and unlevel floors can create turning moments that cause slippage.

## Chapter 4

4.1 Be sure to include an elevation view as well as a plan view of the area scanned by the camera. The resulting video field will be a trapezoid with the "long" edge at the farthest points from the camera. If the camera is a typical 620×480 format, the long edge will be about 15 m long (longer if the camera is pointed farther away so that the near edge is not directly under the camera, or the land slopes downward, away from the camera). This would give a resolution of about 24 mm (an inch), which may not be small enough for mowing accurately. The sloped edges of the trapezoid would have a larger resolution, since the lengths would be about the same, but divided by 480 pixels. The resolution would be improved if a high definition camera was used, but these are not yet standard for a security camera system that fits within our budget. The results will be worse if the camera angles cover more distance.

4.2 Robot-mounted cameras would seek to identify objects in focus from near (about 1 m) to far (about 20 m). As with the previous question, resolutions on the order of 30–35 mm would be typical at a range of 20 m. Each object (perhaps a striped pole) would need to be in a known location in x-y in the field. Since there is no reliable "depth" or distance measurement available with a robot-mounted camera, three such objects would need to be located and pair-wise angles relative to the robot derived from them. The three objects would need to be located at the same time, so the robot speed would need to be negligible in the 33 ms time window of the NTSC video. (If the camera operates in an interlace mode, the time would be halved, but the resolution doubled.) Errors in location would be strongly dependent on the relative angles found, since the calculated location would be confounded by several factors. First, a long distance object location would have errors on the order of 30–35 mm; secondly, short distance object location would have errors due to locating the center of the object in the video field spanning several pixels; and thirdly, the subtended angles computed would depend on the resulting location. This means that the best resolution would occur with the robot in the center of an equilateral triangle, but that the location error would be greater as the subtended angles became smaller or larger. An ideal case would have each subtended angle equal to 90°, but that condition cannot be met with three points, requiring four. This case would entail a relatively large number of objects located in the field.

4.3 Any located point would not necessarily imply a single location in a field of objects, given finite intervals of resolution. Identification of each object would need to be actively tracked. One way to do this would be to trace the robot location within a field map of the entire area, such that objects that move out of range or view could be accounted for. Further, a known starting point and orientation would be needed for the robot in order to avoid any confounding of position computed from a pair-wise (or better) calculation.

4.4 A simulation will show that the random plan will asymptotically approach full coverage in a finite time, but that asymptotic curve will depend on the ratio of the field area and the width of the path. The user has a choice in running the random machine until its batteries lose charge, or until the missed places are small enough to be ignored. The student should write a simulation (in any computer language) to find the relationship sought using a circular pattern for the mower and a number of ratios between the field size and the mower cut radius. In this way the azimuth of the mower can be ignored. Also the angle of incidence should be increased to compute the angle of reflection at a wall to avoid limit cycles, and more closely approximate the slippage of an actual mower.

4.5 The allocation would be the sum of the costs of each subfunction in the next level of detail. Similarly, the next higher level allocation would be the sum of *Cut Grass* and *Plan Path(s)*. Specifically, *Select Waypoints* and *Map Area* are presumed to be accomplished by the user.

4.6 The ruling relationships are that a sound wave loses power proportionally with the square of the distance to the sensed object, and the returning beam similarly reduces as the square. Thus, the function of sound wave reduction is the fourth power of the distance, at best. Mitigating this fact even further is that 55 kHz sound waves attenuate much more than that, due to absorption in the air. The 55 kHz used by the Polaroid® system is actually a mixture of frequencies between 50 and 55 kHz since some targets absorb less sound than others. For example, a fur coat becomes very difficult to detect, but a pair of glasses reflects sound very well. The gain versus time plot should be a straight line on log-log paper with a slope of 4 or more to maintain the same amplitude of received signal. The limitations on this graph would consider the noise level threshold of the detector and can be found by just a few measurements. The actual circuitry in the camera-based SODAR system employ an exponential curve for gain that starts at unity and rises to about 1 million in the space of about 40 ms. At the end of this time period the lens focus is already at "infinity," so this auto-focus feature is applicable to near distances only. This works well for most pictures, since users tend to take pictures of other people nearby or very distant scenes.

4.7 The gain-variation keeps the relative noise level lowest, so that loud sounds do not confound the lens adjustment function. If extraneous noise was not considered likely, then no variation would be needed at all. A send–receive sound system at 55 kHz would operate over a very limited range and would not be suitable for our application. Any send–receive system is based on echoes, and multiple echoes are known to be very difficult to distinguish from the principal distance measurement we wish to find.

4.8 A tag line is a light-weight cable wound around a pulley that drives a multi-turn potentiometer. Direct distance measurements at high accuracy are possible between a robot and a fixed attachment point. A possible problem occurs if the field to be mown has either any objects in it (trees, bushes), or is hilly. In these cases the light cable may snag and give false readings to the robot controller. Also, the potentiometer attachment would need to account for the azimuth rotation of the robot: if the potentiometer was fixed, it would need to rotate about a single point with little or no friction; if the potentiometer was attached to the robot, it would need to rotate with little or no friction continuously, suggesting the need for "slip-rings" to conduct the signal. These are typically quite expensive for high-quality signal conduction. Finally, at least two such tag lines would be needed to triangulate to a field location. Even this method returns ambiguous results since any two distances would compute a pair of positions with respect to the fixed points.

4.9 Given that the wheels are new and clean, the skid-steering method could deposit a bit of rubber on the floor with every turn made. Such an arrangement would also potentially put strong lateral shearing forces on a rug, causing it to wrinkle. Finally, the localization method employed may need additional coding to account for objects in the "field," such as chairs and tables. This problem is avoided in the outdoors by planning the paths as shown in Figure 4.2, and using discrete "patches" to be mown consecutively. Figure 5.24 illustrates how this approach avoids obstacles by path planning.

4.10 This answer is to be determined by the instructor and student together.

4.11 The three methods are: rotary mower, reel mower, and string trimmer. Rotary mowers are far less efficient than reel mowers or string trimmers due to the relatively large amount of air movement they produce. Commercial "random pattern" mowers improve this inefficiency by using short blades with small angles of attack compared with a common gasoline-powered rotary mower. This air movement is an advantage in helping to move the clippings into neat rows for later collection, or to have them ducted into collection bags. The rotary mower needs to keep its drive wheels clear of the spinning blade, so a three-point support is often used, that is, two drive wheels and a single trailing caster wheel. In this way it becomes sensitive to small dips and bumps on the lawn.

The reel mower needs to free of debris, since the cutting blades shear the lawn against a fixed knife blade. By moving little air, it cuts efficiently. Its inefficiency derives from the adjustment of the fixed knife blade to the reel blades. This adjustment sensitively affects friction and cutting quality. Therefore the reel mower cannot be used to clear the lawn of leaves in the fall. Collecting clippings is routine, however, with an open-toped bag at the rear of the mower.

The string trimmer relies on a high tip speed to tear the tops off grass blades or to shear the stems of weeds. String length needs to be maintained regularly for wear. No clippings collection means are included, but the system is very robust when encountering debris. If supported by wheels, as in a robot, it similarly needs to keep its drive wheels clear of the spinning string.

## Chapter 5

5.1 For an op-amp used as shown in Figure 5.6, the rule for the overall gain of the circuit is: gain = 1 + Rf/Ri. This can be shown by considering the input current to the op-amp is nearly zero and the open = loop gain is very large (a million or more). Since Rf and Ri are equal, the overall gain is 2.

5.2 Checking the schematic diagram for the internal functions of the CD4013 reveals that any clock pulse transfers the D input to the Q output, and that Q-bar is always the opposite of Q. Connecting the D (data) pin to the Q-bar (inverted output pin) will cause a zero to change to a one upon receipt of a clock pulse, and vice versa. The reason that the CD4013 is present is to solve the problem of the Sync Out pulse being too short to be reliably captured by the #1047 board. Dividing by two allows the #1047 to detect the change of state at any time.

5.3 A logical circuit can be devised using one-shot timers.

5.4 A simple oscillator formed with a CD40106 is exquisitely sensitive to applied voltage. A 10% change in supply voltage could likely change the sound sender frequency enough to reduce the effective range greatly. A well-regulated supply is needed (see Section 5.2).

5.5 Channel B is in quadrature with channel A, and is shown in Figure 5.16.

5.6 Assuming that the time of flight of the sound signal is 335 m/s, we read from the screen 17.5 and 14 ms, respectively, for 5.86 and 4.70 m.

5.7 There are two main reasons: the `atan2` function accepts a value of zero for Δx; and the signs of the values for `Δy, Δx` are preserved so that the angle appears in the correct quadrant.

5.8 A rectangular patch that is 72 in. × 570 in. along the centerline of the mower, so we need to add about 18 in. to the width giving $90 \times 570 = 51{,}300$ in$^2$, or about 33 m$^2$.

5.9 Many ground-based R/C makers have switched to 2.4 GHz and "crystal-less" designs for transmitters and receivers. No functional changes would be needed in specifying the new hardware, just the allocations. If one were to choose a three-channel receiver over a four-channel receiver, channel 3 would take on the role of the previous channel 4. The pulse position modulated (PPM) scheme is still the standard for R/C vehicles.

5.10 The amplifier gain of the receivers in Figure 5.13 show a gain of 22,000 then a "noise gate" followed by a gain of 5.7. The noise gate cuts off any signals below 0.7V. If the received signals rise gradually from zero (or background noise) to a value above the 0.7V at the noise gate, then the circuit may "miss" the first or second pulse in the burst of about 25 that is sent from the transmitter in Figure 5.9. Since each pulse in the burst occurs at 40 kHz, with a period of 25 μs, we could miss 50 μs, or 0.05 ms. This corresponds to about 15 mm of lost distance. Note that the errors will always be less than the actual value.

5.11 With several thousand dollars to allocate, one might consider alternative radio transmitter and receiver devices. Alternatively, an on-board computer would be considered a radical change, in which case functions to communicate with the sound receivers by radio could be designed, changing the wire-links between a master station to fully wireless. An important allocation that cannot be missed is the non-line-of-sight communication distance of about 1 km, in case the lawn area demands it. Other ideas would be welcome in this open-ended question.

5.12 The principal reason is lack of ground friction in azimuth changes. This has the dual entailment of high sensitivity to uncommanded turning due to dips and bumps on the lawn.

## Chapter 6

6.1 Slippage could increase with wet grass or mud. The control system relies on feedback from the sound receivers, so the long-term robot position may be realized, but the speed with which it could respond may become highly variable. The robot becomes uncontrollable when the slippage cannot be corrected by the feedback loop of Figure 6.29. That controller makes no assumptions about disturbing conditions caused by slippage.

6.2 The function of the LM567 is to detect tones of a given pitch. Without it, one could realize a phase-locked loop design digitally, or resort to an analog band-pass filter with a high "Q" followed by a 20 ms "one-shot."

6.3 A hands-on problem that is answered by experiment, since a wide range of "k" values exist for small DC motors.

6.4 Bouncing is an indication of too great a forward speed. A reel-type mower needs a minimum forward speed to operate well, however. One of many ideas: use the robot to draw a heavy lawn roller to even the turf. No minimum speed required. Another idea: search for the cause of the unevenness (such as mole hills) and eliminate it. As a last resort, one could consider other types of mowers, such as the inefficient but more forgiving string trimmer design. Note that little would change in the function diagram regarding control and navigation, but a significant design problem would need to be solved. An alternative, but not recommended due to complexity, could be a four-wheel drive system with a spring suspension. Special attention would be needed to ensure a constant, desired cutting height.

6.5 The functions at the right-most "column" could be rearranged to approximate a flow chart of a generic sequence of operations to be performed. That sequence would be informed by a study of the software in Chapter 7, especially the order of the functions called in `waypoint_drive()`.

6.6 The keyboard code for lower-case "a" is the number 97. This number is derived from long-standing ASCII-7 codes used before the invention of PCs.

6.7 Many such functions operate "in background mode" at all times to detect data from the sound receivers. For example: __stdcall() and its subfunctions.

6.8 The "ground" of Figure 5.9 is also the center (white) wire between the two main batteries. Figure 6.5 shows the ground shifted to the minus pole of the "lower" battery. Still, the op-amp of Figure 5.9 needs ±20 V to send a loud signal. These two voltages are generated by DC/DC converters that extend the +24 V of Figure 6.5 to +32 V, and the ground of Figure 3.7 to −8 V. The difference between +32 and −8 is, of course the 40 V needed, with the white +12 V as the ground.

6.9 PPM strings of four timed commands are transmitted to an R/C receiver. This receiver separates each tandem pulse into four streams that repeat every 20 ms. The first two of these streams are used to drive a power driver (in this example a Scorpion™). The output of the power driver is a high current, pulse width modulated(PWM) PAIR of currents that have varying widths repeated at 50-μs intervals. Each width is determined by the width of the incoming PPM pulse in a specific way: a 1 ms width commands the minimum width of the PWM current, while a 2 ms width commands the maximum width of the PWM current. It will be seen that the 20 ms PPM repetitions do NOT correspond to the 50-μs PWM repetitions.

6.10 The file "`stack [61] [6]`" is a push-down stack (most recent time is the first set of six values) that is used by the "`file_write`" function called by `waypoint_drive`.

6.11 One would need to sense the stall condition, for example, by noting three types of changes from the planned path: sharp right turn, sharp left turn, no motion forward at all. Then one would drive both drive motors in reverse for a short time. Finally, the routine would return to the calling function.

6.12 Classical low-pass filters work for Gaussian noise. The simplest of these is passive, consisting of a series resistor followed by a parallel capacitor. Such a passive filter drops the forward gain of the system in which it is used. For active low-pass filters, please refer to the *Active Filter Cookbook* by Don Lancaster [1].

6.13 There are four cases to consider: a chain drive with either an even or an odd number of links, and sprocket wheels with either an even or an odd number of teeth. The hole spacing must be a multiple of one quarter of the chain pitch depending on the design selected. For example, an even number of links with an even number of teeth require a hole spacing with a multiple of one chain pitch.

6.14 One would extend Figure 6.21 to another level of functionality at *Skid Steer,* substituting a different function, such as *Caster Steer.* Note that both initial words are used as verbs, not adjectives. Figure 6.9 may also be affected since the artifacts of Side Brackets and Side Frames would change. The number of drive wheels can be expressed as an allocation, as well as the number of casters, which could be any number, but at least one. Note that commercial lawn tractors often employ two drive wheels (under the operator) and two casters (at the extreme forward positions.)

6.15 Mecanum wheels rely on reliable and constant friction between the wheels and the ground, plus a relatively flat floor so that all four wheels are engaged at the same time. Skid steer, while less efficient, can tolerate rougher surfaces and variable friction.

6.16 If one noticed that the robot tended to wander from the planned line, one would suspect that `d2line` had almost no affect at all on `delta`. Changing `kp` could be accomplished most effectively by simulating the robot control loop in Figure 6.29 and deriving a root-locus plot [2]. The effects of kp would then be directly seen on the locus, and one would select values that placed the closed-loop poles near the line extending from the origin at 45° to the left. Alternatively, one could double the present value of kp and run a test to see if the wandering persisted. If the wandering became very arcuate, one would know that at least one of the values kp, kd, or ka was too high.

6.17 At the scale of the servo driver, one would see from mov[i] if the commands to the motors were oscillating or remaining steady. This could be a first indication that the servo loop of Figure 6.29 may have inappropriate values. Useful to the operator if the robot seemed to be somewhat non-responsive to the programmed path.

6.18 The values of all the pertinent control variables are displayed and saved in a list. Each of these should tend towards zero for optimum operations. If delta, for example, were always high and saturating at the programmed limit, then one would be able to see this condition and the state of the other loop variables to determine which might be changed.

6.19 All function declarations can be eliminated by checking that all functions appear in the listing in the order of being called. The application programmer interface declarations (for the Phidget Cards) are always needed and are usually placed at the top of the globals list.

6.20 The integrals accumulate in value, without limits, except due to saturation in practice. The premise of Figure 6.29 is that the robot control remains in its linear range and does not saturate. If not otherwise compensated for, the robot may travel on a fixed arc until the linear range is again met.

6.21 Adding up the differences between each wheel motor results in a model of the heading, less a fixed starting point (the constant term of any indefinite integral). The effective starting point is quickly found by computing the vector of actual positions of the robot using the distances from the Sound Sensors.

6.22 The average, or mean, is valid when the data form a Gaussian, or normal, distribution. If the actual data do not conform to the normal distribution, then the mode will save the most-often found value, which may well apply to systems with digital components in their calculation.

6.23 As the robot approached the end point, it would tend to steer towards it, resulting in an unplanned angular orientation with each waypoint, at best. In the worst case, the robot could miss the end-point completely as it attempted to spiral inwards towards it.

6.24 The robot would enter a zone in which the triangulation would fail, and the net motion could be along an unplanned vector. When the triangulation began to function again, the robot would likely steer "back" to the planned vector.

6.25 From left to right, the two receiver cables each carry power, ground, and signal, in form of pulses as seen in Figure 5.21. The main transmitter cable carries only power, ground, and signal, in form of pulses as seen in Figure 6.8. The next three cables are USB standard cables carrying power, ground, and two signal lines each. Finally, the cable on the far right is another USB cable connecting a mouse to the control computer.

## Chapter 7

7.1 Generally good programming practice would suggest replacing subscripts with pointers. The present code was not written for speed, but for easy comprehension. If there were no subscripts involved in the offending function, one would look to the functions that it calls. Alternatively, one could invoke the slow function on alternate cycles of 160 ms.

7.2 Two time-critical code loops exist in the code: the first is invoked by interrupt and is not affected by `curses.h`. The other is the `waypoint_drive` function that controls the robot motors. That function could include a `getch()` for the span of time remaining in that function to total 100 ms.

7.3 Rather than moving the data in the stack, one could change just a pointer to the values corresponding to the present time. The "pointer" could be a simple index to the current stack location, while the stack became populated by new values one line at a time.

7.4 The fourth channel is always used as a marker to indicate that a synchronizing event has taken place. This marker avoids the use of a fixed time clock and its attendant drift over time. The fourth channel contains no other information since its length is fixed. The synchronizing event causes a function call to `__stdcall InputHandlerE()`, which computes the distance to each receiver from the robot's sound sender.

7.5 New functions would include a new call to read the value in channel 3, in the same way as `InputHandlerE()`, finds the time signal in channels 1 and 2. The value updated would need to be expressed as a global, and stored in `mov[2]`.

7.6 For the student to explore.

7.7 One could, for example, employ a low-pass filter in several guises. Entailments would include a change in amplitude for the signal and a shift in time (phase) due to the filtering. Another possibility would be to convert the digital data to analog (with a DAC) and use a filter network known to feature a constant gain, and then reconvert to digital with an ADC [1].

7.8 The student should include the formulas for solving general triangles with the side-side-side case. A generic triangle with three sides and three angles needs to be shown, which correspond to the fixed distance between sound receivers and the distance found by `InputHandlerE`. In addition, special cases for unsolvable triangles need to be trapped in two ways: by failure to "close" and by failure of the computed arcsine to be in bounds. Portions of the code that respond to such errors need to be highlighted.

7.9 When a "sync" event is sensed from the radio receiver (Figure 5.9), the corresponding encoder receives an "Index" pulse which resets a counter in the encoder to zero. Subsequent pulses on the "A" and "B" pins count up at the rate given by the clock input (also Figure 5.17) until a sound pulse is received which stops the clock signal. Figure 5.15 is not especially sensitive to its supply voltage (from a USB source), but the clock frequency IS. Should the clock supply voltage change, the clock frequency will also change in a rapid and non-linear way. Please refer to the CD40106 data sheet for details; see Appendix D.

7.10 When the distance between waypoints is reduced to zero, the denominator of the `dist2line()` function vanishes, causing a "divide by zero" error.

7.11 The slope of the computed line is set to 1000 rather than a value that may cause a floating-point error. The lateral distance error is limited to 1/1000 of the waypoint's separation.

7.12 This function seeks the statistical mode of a pair of values from a series of values found by the interrupt routine `InputHandlerE`. This mode represents the best value of distance that can be measured given that there may be jitter in the measurement. The routine works by taking readings of the distances every 80 ms so that data repetitions will not occur. For each of two distances it collects a list of values in "raw" form, that is, before being converted into inches. If a reading matches a previous one, a counter (list[][]) is incremented to build up a histogram. If a sufficient number of matches are found (20), the routine ends. It finds the peak number of occurrences and retrieves that pair as the modes. It converts the mode values to inches and returns them in avg[0] and avg[1]. It ignores data dropouts and zero values. This routine is called by locate() only if the robot is not moving. It is used to help locate the receiver stations and any other points of interest in the workspace of the robot. Since it uses many instances of distance obtained on an 80-ms cycle, a substantial fraction of a second is needed for completion. While superior in removing jitter in the sound pulse durations received, it would not supply timely updates to waypoint_drive.

7.13 The "variable" sx is really a constant if n is constant, also most of the numbers could be stored in a short list using a pointer rather than a subscript. A few multiplies and divides could be re-grouped for faster execution.

7.14 The operations would be affected in at least two ways: first, the robot would not jerk to speed or to a stop; and secondly, the actual distance traveled to a waypoint would be lengthened by the "coast down" phase of the commanded motion. If the feedback parameters were wrong and led to erratic operation, the Acceleration and Ramping would tend to smooth-out the motions, but not the commands.

7.15 Yes, any abrupt velocity changes would be smoothed out.

7.16 The robot may under- or over-shoot the region of non-computable triangulation causing a sideways "jog" in the motion.

7.17 One would need to include a list of historical points to which a curve could be fit. The parabola is the simplest with a closed-form answer. The circle is closed-form for only three total points, and best-fit by iteration for more points.

7.18 The answer is under ALL conditions, since its distance is not compensated for in waypoint_drive. This is especially apparent when the distance between waypoints diminishes to about a meter.

7.19 Some ideas: look to the stack of xe, ye, ze positions for an imputed position that fits the current xe, ye, ze values found from sound sensor feedback; implement a "back-up" motion and then re-establish a position; interrupt waypoint_drive and let the operator take over manually for a correcting move, and so on.

7.20 The last line reports the average magnitude of the variable d2line , indicating the magnitude of the expected error from the planned path.

7.21 The two lines that check for errant values could be eliminated since atan2 is defined to return well-defined values.

7.22 When stnd_dev() is called with n = 1, which is not checked.

7.23 The Linker must reference the library phidgets21.h.

## Notes

1. Lancaster, D. (1996) *Active Filter Cookbook*, 2nd edn, Newness.
2. Phillips, C.L. and Harbor, R.D. (1988) *Feedback Control Systems*, Prentice Hall.

# Index


align motors, 90
allocation, 3, 21, 27, 33, 38, 43, 45, 65, 79, 80, 84
amplify signal, 47, 53
apply torques, 12, 17
"as low as reasonably achievable" (ALARA), 12, 19
assemble arm, 9, 11, 15

bandpass filter, 52
base board, 84, 85
battery strap, 83, 85
beep, 59, 61, 118, 124, 126

16C54, 98, 99
center signal, 47, 52
channel head, 7, 9, 10, 12, 13, 19, 20
coal mine, 23, 25
coal scoop vehicles, 25–7
conceptual design, 5, 12, 21, 38, 40, 42, 45
conduct current, 75, 79
connect controls, 11, 15
connect signals, 56, 57
contact ground, 89, 90
contact obstacle, 90, 92
continuous coal haulage, 25
continuous miner, 2, 25–7, 31
control access, 27, 28
control block diagram, 94, 95
control heat, 27, 29, 31
control positions, 12, 17
control system, 16, 31, 36, 42, 73–6, 79, 93
control tools, 2
control tracks, 28, 30, 31
convert pulses, 53
convert signals, 47
convey coal, 27, 28
cross beam, 89, 91
crosstalk, 53–5
curses.h, 75, 76, 79, 98, 107–9, 136
cut grass, 36, 38, 40, 43

debug functionality, 56, 57, 62
design infrastructure, 74, 76, 88
determine kinematics, 27–9
differential GPS (DGPS), 39
display, 18, 21, 56, 62, 79, 96–9, 108, 122, 126, 130, 136
display_Eproperties, 77, 78, 97, 98, 135
display gaps, 56, 62
display_Sproperties, 77, 97, 98, 136
display sync, 56, 62
dist2line, 96, 100, 111, 136
divide tasks, 2



*Practical Field Robotics: A Systems Approach*, First Edition. Robert H. Sturges, Jr.



Companion Website: www.wiley.com/go/sturges

drive MBCs, 27, 28
drive wheels, 85, 89, 90, 126

enable power, 75, 79
encoders, 35, 58, 60, 95
ErrorHandlerE, 100, 130
ErrorHandlerS, 100, 130
establish geometry, 74, 93, 100
estimate cost, 45, 46
evaluate complexity, 70
evaluate cost, 70
expand pins, 12, 16
expedite inspection, 2, 11
extrapolate, 33, 56, 59, 62, 64, 65, 100, 124, 137

feedback position, 11, 14
fileread, 77, 96, 100, 121
filewrite, 94, 96, 125, 133, 137
find accuracy, 45, 46
find_diff, 115
find_dir, 94, 96, 114
find_line, 59, 61, 62, 94–6, 113, 136, 137
find location, 36
find_point, 95, 114
find range, 45, 46, 68
find_ze, 116
follow plan, 27
follow rules, 27
front panel, 83, 85, 98
function block diagram (FBD), 3, 11, 26, 33, 96

generate quadrature, 56, 58
generate sync, 47
get_heading, 94–6, 119, 137
get_intersection, 100, 101, 112
get_jstar, 100, 101, 111, 112, 123, 137
get_line, 100, 101, 111, 136
get_median, 94
get_mode, 77, 96, 97, 112, 124, 137
get_perp, 100, 112, 136
get_ze, 114, 116
graphics user interface (GUI), 76
guide arm, 11, 14
guide un/coupling, 11, 12, 14, 15, 17
guide walker, 11, 14

handle dropouts, 56, 62
handle exceptions, 74, 93, 100
hauling, 23

inertial guidance, 39, 40
inform user, 74, 93, 97, 136
InputHandlerE, 56, 59, 60, 96, 97, 100, 110, 112–14, 119, 123, 131, 136
insert pins, 12, 16
inspect tubesheet, 2, 12
install/remove arm, 11, 13
install/remove tools, 2, 11, 13, 17
install/remove track, 11, 13
install/remove walker, 11, 13, 14
interference, 52
interpret signals, 56
interpret sound, 36, 56–8, 62
interrupt, 60, 61, 96, 110, 112, 113, 123, 125, 130, 131
iposition, 59, 61
isolate weather, 42, 66, 67

laser speed measurement (LIDAR) devices, 39
limit crosstalk, 53, 55
limit noise, 53, 54
limit voltage, 56
link conveyors, 28
link ends, 27
LM317, 51
LM358, 50, 72
LM567, 81, 82, 103
LM675, 52
LMM3 code, 46, 47, 57, 58
localization, 8, 39, 40, 78
locate 9, 17, 21, 76, 77, 79, 92, 96, 97, 101, 112–18, 120, 121, 123, 124, 126, 128, 132
locate conveyors, 27, 28
locate robot, 74, 93, 94
long-wall, 23, 25, 26, 31
lost 124, 125, 131, 137

main 47, 60, 61, 75–9, 103, 107, 121, 122, 128
main receiver, 47, 51, 75, 81, 82, 98, 136
main transmitter, 47–9, 50, 62, 81, 102
map area, 36
master, 12, 16–18, 20
maximize safety, 28
maximize throughput, 27
measure extension, 28–30
meet schedule, 2, 3, 11
microphones, 53, 54
Microswitches™, 90
mine coal, 27
minimize forces, 27–9

minimize weight, 2, 11, 12
mobile bridge conveyor (MBC), 27–31
model mode, 17, 19
modulate speeds, 75, 82
monitor slippage, 28, 30
motor gears, 88, 90
motor mounts, 89–91
move master, 12
move robot, 74, 89, 90
move_servo, 75, 77, 121, 122
*Mow lawn*, 34, 36, 43, 51, 73, 78, 89, 101, 103, 105, 120, 121

navigate containment, 2, 12
navigate tubesheet, 11, 13
navigation and mowing, 34
noise gate, 55
nuclear service, 2, 7, 9, 11, 13, 15, 17, 19, 21

odometry, 39
open manway, 11, 13
open_encoders, 60, 77, 78, 97, 98, 134, 137
open_servos, 77, 79, 97, 98, 135, 137
operate miner, 27
operate robot, 36, 38, 42, 73, 75, 93, 99
operate track, 11, 13
oscilloscope, 56, 63, 102
override detector, 75, 81, 82
override receiver, 75, 81, 84
override transmitter, 50, 81, 102

phidget21.h, 73, 74, 126
pillars, 24–6
plan path, 36–8, 91, 92
plan process, 2, 11
position sprockets, 90
power logic, 47
power receivers, 47, 53
power subsystems, 74–6, 79, 80, 103
power transmitter, 47
prevent leakage, 2, 3, 11, 12
principal, 17, 42, 72, 79, 94, 96, 111, 115, 119, 133
program speeds, 74, 88
protect circuits, 75, 79
pulse width modulation (PWM), 49, 75, 81, 82, 95, 96, 121, 126
push, 59, 94–6, 131
push4, 94, 96, 132
pushdown, 94–6, 132
pushflag, 59, 94–6, 131
PWM controller, 75, 82, 95, 96

quadrature, 56, 58

radio control (R/C), 48, 50, 51, 69, 73, 74, 81, 82, 84
radio link, 40, 42, 47, 68, 73
ranging, 38–40, 43
reach tubesheet, 2, 11, 13
receiver board, 53–5
receivers_in, 77, 79, 100, 108, 109
receivers_out, 77, 79, 100, 109
receive signals, 36, 42, 73–5, 82
receive sound, 36, 40, 48, 53, 62, 63
receive speeds, 75
receive sync, 47, 50
repair tubes, 2, 12
rooms, 24, 25
rotate body, 11, 15
rotate motors, 74–6, 81, 82

safety and reliability, 19, 30, 37
save history, 74, 92–4, 96
screen_end, 78, 97, 98, 109
screen_init, 77, 97, 98, 107, 109
sealed lead acid (SLA), 79
segment arm, 2, 21
select environment, 66, 67
select frequencies, 36, 48, 65, 68
select language, 66
select motions, 36, 65, 70
select mower, 66
select platform, 36, 66
select swath, 36
select waypoints, 36
send pulses, 47
send sound, 36, 46, 47, 62, 80
sense obstacle, 90, 92
sense personnel, 27, 28
sense walls, 28–30
sequence motions, 74, 92, 93
service arm, 20, 21
service exchanger, 2, 3, 11, 15
servo board, 46, 47, 50, 73, 79, 81, 130, 136
servo motor, 48
servo_go, 75, 77, 80, 81, 122
set channel4, 47
set parameters, 36, 38, 40, 65, 79
side brackets, 83, 86

skid steer, 70, 89, 90, 96
smoothed_dist, 59, 61, 95
smoother, 94, 96, 97, 108, 117
sound detection and ranging (SODAR), 39, 40, 43
spacer bars, 87, 88, 91
startout, 117, 121, 124, 132, 137
start timing, 56, 60
stnd_dev, 95, 97, 98, 114, 132, 137
stop robot, 80, 89, 90
stop_servo, 75, 77, 80, 122, 128
stop timing, 56
suburban lawns, 33
support conveyors, 27
support walker, 11, 12, 15, 16
survey technologies, 36, 46
switch, 60, 71, 75, 77, 79, 80, 92, 128
synchronize wheels, 90

tag line, 39, 40, 43
teleoperation, 2, 3, 19, 22, 42, 47
teleoperators, 42
test_servo, 75, 78, 80, 122
theoretical robotics, 1
time_reads, 128, 136
transducers, 47, 52–5, 69
translate signals, 56, 58
transmit sync, 47
triangulate, 62, 65, 94–7, 101, 111, 112, 118–20, 123, 124, 126, 136

ultrasonic, 39
underground mine, 2
underground mining, 23
Universal Serial Bus (USB), 46–8, 50, 51, 57, 59, 60, 61, 67, 73, 74, 78, 81, 98, 100–102, 110, 121, 128, 134, 135, 136
update_display, 98
update_display2, 108
update_display3, 108
update_display4, 109
update_display_servo, 108
USB Translator, 50, 51, 57, 78, 81, 101, 102, 110, 136

value analysis, 3, 4, 31, 103
value engineering, 3, 31, 36, 46
Velcro™, 90

walker, 9–21
waypoint_drive, 78, 90–93, 96, 101, 111, 115, 117, 120–124, 126, 128, 131–133, 137
waypoints, 36, 38–40, 64, 91–3, 95, 96, 101, 111, 120, 124, 125, 132
wheels, 70–72, 83, 85, 86–91, 103, 126
while, 77, 79

xyzwrite, 134

Printed in the United States
By Bookmasters